National
Fire Protection
Association

LADDER COMPANY
FIREGROUND
OPERATIONS

HAROLD RICHMAN
Fire Chief (Retired)
Past President
International Society of Fire Service Instructors

JONES AND BARTLETT PUBLISHERS
Sudbury, Massachusetts
BOSTON   TORONTO   LONDON   SINGAPORE

Jones and Bartlett Publishers

World Headquarters
40 Tall Pine Drive
Sudbury, MA 01776
978-443-5000
info@jbpub.com
www.jbpub.com

Jones and Bartlett Publishers Canada
6339 Ormindale Way
Mississauga, Ontario L5V 1J2
Canada

Jones and Bartlett Publishers International
Barb House, Barb Mews
London W6 7PA
United Kingdom

National Fire Protection Association

1 Batterymarch Park
Quincy, MA 02169-7471
www.NFPA.org

Jones and Bartlett's books and products are available through most bookstores and online booksellers. To contact Jones and Bartlett Publishers directly, call 800-832-0034, fax 978-443-8000, or visit our website www.jbpub.com.

Substantial discounts on bulk quantities of Jones and Bartlett's publications are available to corporations, professional associations, and other qualified organizations. For details and specific discount information, contact the special sales department at Jones and Bartlett via the above contact information or send an email to specialsales@jbpub.com.

Production Credits

Chief Executive Officer: Clayton E. Jones
Chief Operating Officer: Donald W. Jones, Jr.
President, Higher Education and Professional Publishing: Robert W. Holland, Jr.
V.P., Sales and Marketing: William J. Kane
V.P., Production and Design: Anne Spencer
V.P., Manufacturing and Inventory Control: Therese Connell
Publisher, Public Safety Group: Kimberly Brophy
Senior Acquisition Editor: William Larkin
Editor: Jennifer S. Kling

Production Supervisor: Jenny Corriveau
Production Editor: Karen Ferreira
Photo Research Manager/Photographer: Kimberly Potvin
Director of Marketing: Alisha Weisman
Front and Back Cover Images: © Glen E. Ellman
Composition: OmegaType, Inc.
Illustration: Electronic Illustrators Group
Text Printing and Binding: Malloy
Cover Printing: Malloy

Library of Congress Cataloging-in-Publication Data
Persson, Steve.
  Ladder company fireground operations / Steve Persson, Harold Richman.—3rd ed.
    p. cm.
  Rev. ed. of: Truck company fireground operations / Harold Richman. c1986.
  Includes index.
  ISBN-13: 978-0-7637-4496-0
  ISBN-10: 0-7637-4496-4
  1. Fire extinction. 2. Fire engines. 3. Aerial ladders. 4. Ladder companies. I. Richman, Harold. Truck company fireground operations. II. Title.
  TH9310.5.R53 2007
  628.9'25—dc22
                              2006037061
6048

Printed in the United States of America
12 11 10 09 08   10 9 8 7 6 5 4 3 2

Harold Richman
November 9, 1926–October 16, 1999

A TRIBUTE TO THE AUTHOR AND HIS LEGACY

John R. Leahy, Jr.

Hal started his legacy under unusual circumstances as a volunteer with the "War Time Preparedness Civil Defense Program" in our nation's capital. The number one priority was engine and truck company operations, training under suspected air raid activities similar to the British Fire Brigade.

When the opportunity arose, he joined the United States Marine Corp until the end of the war. He became a member of the newly formed, largest all-civilian fire department of the Federal Civil Service—the 14th District Pearl Harbor Fire Department with 21 fire companies, three crash crews, and two marine fire boats with Hal serving as a Captain, Training Officer, and District Chief.

Eventually Hal returned to the mainland as a fire fighter in Memphis, Tennessee. He was promoted to training officer and then accepted the challenge to expand recruit training. The need was to fill the personnel numbers with people exposed to the latest technology and skills, so he reviewed programs across the country to put in place a quality program. He helped manage these changes as a Shift Commander and implemented a rescue company operation. Subsequent to that, Silver Springs, Maryland, convinced Hal to take over their training and operations program for standardization of heavy commercial and residential growth.

Hal's life was built upon the fundamentals of training; and he was constantly preaching, to anyone who would listen, from the basics to the university level.

He eventually accepted the first career fire chief position of Fairfax County, Virginia, where his concern was for excellence in performance, based on training. Here he was able to implement his dreams one step at a time. From there he went on to become Chief of Prince George County, Maryland, qualifying over 500 officers in the service at the University of Maryland Fire Officer Staff and Command School. Hal also served through the chairs of the International Society of Fire Service Instructors, the National Fire Protection Association, and others; and authored two books that successfully reduced injuries, loss of life, property, and environment. He was a true example of talking the talk, and walking the walk.

Hal was one of the great ones who influenced all those he came in contact with and who inspired them to feel they should influence others as well. I'm proud to say I really knew him personally, and I hope I have done him justice.

Brief Contents

Contents

The National Fire Protection Association and Jones and Bartlett Publishers are pleased to bring you *Ladder Company Fireground Operations, Third Edition*. This new edition, previously called *Truck Company Fireground Operations, Second Edition,* combines current content with dynamic features for both the instructor and the student.

Fire department ladder companies face many challenges. Ladder company personnel are an integral part of firefighting operations at the fire ground, and this book emphasizes the point that fire fighters performing ladder company tasks must be properly trained, possess the proper equipment, and be adequately staffed.

Ladder Company Fireground Operations, Third Edition covers the basic objectives of ladder company work including the assignments of conducting a primary search, rescuing victims, forcing entry, and conducting proper ventilation techniques. This book also emphasizes other areas of importance including pre-incident planning, using standard operating guidelines (SOPs), and working within an Incident Management System (IMS).

The *Third Edition* will help reinforce and expand on the essential information and make information retrieval a snap by utilizing the following features:

Learning Objectives
Identify what students should take away from each chapter.

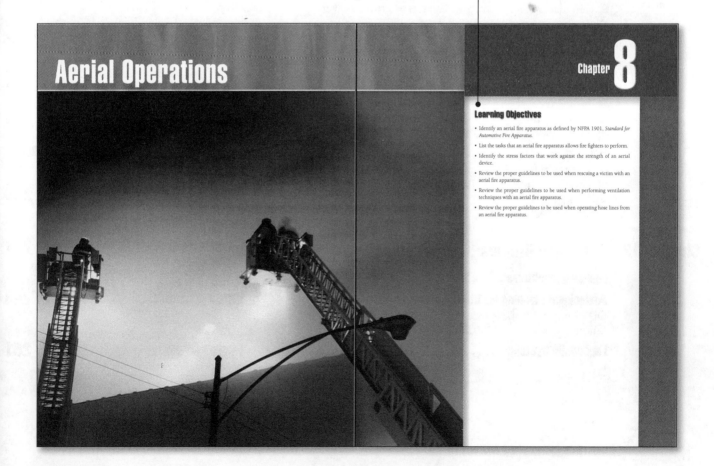

Aerial Operations

Chapter **8**

Learning Objectives

- Identify an aerial fire apparatus as defined by NFPA 1901, *Standard for Automotive Fire Apparatus.*
- List the tasks that an aerial fire apparatus allows fire fighters to perform.
- Identify the stress factors that work against the strength of an aerial device.
- Review the proper guidelines to be used when rescuing a victim with an aerial fire apparatus.
- Review the proper guidelines to be used when performing ventilation techniques with an aerial fire apparatus.
- Review the proper guidelines to be used when operating hose lines from an aerial fire apparatus.

Overhaul operations conducted after the control of a fire incident can be a dangerous assignment for fire fighters. Those who were engaged in the control and extinguishment of the fire may be needed to perform overhaul activities. They may be tired and exhausted from strenuous physical activity put forth while in the performance of their duties during fire suppression operations. Before fire fighters are assigned to overhaul operations, they should be rested or additional personnel should be assigned before overhaul begins. A preinspection should be made to determine the overall condition of the building or area in which overhaul operations will take place. Hazards must be identified and corrections made to ensure the safety of the fire fighters. Command should evaluate the department's risks-versus-benefits guideline before committing fire fighters to an excessively hostile environment.

Fire fighters must be strictly supervised while working inside the building. Crew members must work in teams under the direct supervision of a company or sector officer. Fire fighters must be in full personal protective equipment, including SCBA, as carbon monoxide levels may be extremely high because of the smoldering effects of combustible materials found during overhaul. Safety officers also should be assigned during overhaul to monitor any hazardous conditions and to ensure that fire fighters are working safely. A **rapid intervention team** must be assigned in case an emergency situation occurs.

Overhaul is dangerous work, especially if the building has been damaged excessively. Those performing overhaul operations should pay particular attention to their surroundings and be in direct contact with command. Despite the efforts of the fire service, injury rates remain high during overhaul operations.

When the main body of fire has been brought under control, overhaul operations should commence. Overhaul is the hard, dirty job of searching for and extinguishing any remaining fire, sparks, or embers that are visible or in concealed spaces. The main purpose of overhaul is to make certain that no trace of fire remains to rekindle after fire fighters have left the scene. In addition, overhaul should ensure that the structure is left in a secure state, especially if it is to be partially or completely occupied soon after the fire. Because water damage can occur during overhaul, fire fighters must protect undamaged goods and furnishings as part of the property conservation efforts.

Command must consider the degree of cleanup operations of the premises. Debris should be removed, ensuring total extinguishment of the fire. Command must consider areas of

Key Points

> The main purpose of overhaul is to make certain that no trace of fire remains to rekindle after fire fighters have left the scene.

Key Points

> Debris that could be vital in determining cause and origin should not be removed until an assessment of the area has been made.

the building that may need to be examined by a fire investigator. In this instance, debris that could be vital in determining cause, and origin should not be removed until an assessment of the area has been made. Debris that could be dangerous to fire fighters or to occupants who may continue to inhabit the building should be examined during overhaul and removed from the building if necessary. Fire fighters should perform overhaul operations to ensure that these tasks have been completed. Depending on the circumstances, building maintenance crews may be able to continue cleanup operations after the fire has been extinguished if command has determined that it is safe to do so.

At many working fires, overhaul may be the toughest firefighting assignment. It requires a knowledge of fire travel and building construction, an expertise in hand and power tools used during overhaul operations, and the stamina and muscle for prolonged periods of hard work immediately after fire control. Unfortunately, in many cases, fire fighters already exhausted from strenuous firefighting operations are immediately assigned to overhaul work.

Tired crews sometimes try to work too quickly and tend to take chances in an effort to get the job finished. This often results in mistakes and in injuries to fire fighters. During fire attack and related operations, ladder company members must work quickly and because the nature of the job may take calculated risks. Overhaul operations begins after the fire is placed under control, and there is less reason to rush or take chances. There is time enough for officers to evaluate the situation and develop an orderly and safe overhaul plan. In spite of this, the injury rate during overhaul operations is relatively high.

Three procedures can ensure the safety of fire fighters and reduce the frequency of injury during overhaul:
- Inspection of the area before initiating overhaul operations
- Assignment of rested or additional personnel to conduct overhaul operations
- Proper supervision of personnel during overhaul operations

These procedures are discussed in the first two sections of this chapter, whereas the remaining section deals with the overhaul operation itself.

Preinspection

An inspection of the building before a concentrated overhaul effort is begun should determine the overall condition of the area and ensure the safety of fire fighters assigned to that area.

Key terms
Key terms and definitions are emphasized throughout the text.

Key Points
Key points are highlighted throughout the chapter to help emphasize important topics.

Wrap-Up
Chief concepts and key terms are provided at the conclusion of each chapter.

Fire Fighter in Action
Provide end-of-chapter questions to help students prepare for exams.

Wrap-Up

Chief Concepts

- Preincident planning and inspection are important parts of forcible-entry operations. Ladder companies must be familiar with the different types of entrances to buildings in their district and with the tools needed to force open these entrances. Fire fighters also should be aware of buildings that would present especially difficult entry problems if they became involved with or exposed to fire. Ladder companies need training to cope with these problems, using specially designed forcible-entry tools if necessary.
- Depending on the fire situation, it might be easiest and fastest to force entry into a building through windows. Above the first floor, consider entry through a window, balcony, or porch by ground ladder or aerial device. Otherwise, doors must be forced or walls must be breached. The most difficult doors to force are those that are made of tempered glass or reinforced metal. Thus, it is important that the apparatus carry a full range of forcible-entry tools. These should include standard hand tools for cutting, prying, and striking to power saws, hydraulic cutting and spreading tools, air-powered cutting tools, and chisels and oxyacetylene torches.

Key Terms

Cutting tools: Tools that are designed to cut into metal or wood.

Fox lock: A device with from two to eight bars that hold the door closed from the inside.

Lock pullers: Such as the K-tool, are designed to remove cylinder locks.

Prying tools: Tools designed to provide a mechanical advantage using leverage to force objects, in most cases, to open or break.

Striking tools: Tools designed to strike other tools or objects such as walls, doors, or floors.

Fire Fighter in Action

1. Forcible-entry operations can
   a. Add damage to the fire building
   b. Allow fire fighters to get into position quickly to reduce overall damage
   c. Allow fire fighters to get into position quickly to save lives
   d. All of the above

2. When first-arriving ladder companies are confronted with a working fire that has gained headway with a possible life hazard, they should
   a. Perform a quick 360-degree walk around, checking for alternative access to the fire building that does not require forcible entry, heeding the adage "try before you pry."
   b. Quickly check the fire building to determine the easiest way to gain entry and limit damage.
   c. Force entry into the fire building immediately without stopping to consider the damage.
   d. Force entry into the fire building immediately, but consider the damage.

3. When confronted with a working structure fire with exposures on either side, ladder crews assigned to forcible entry should
   a. Check the door to be sure it needs to be pried and also check with management, occupants, or security to see whether they can open the door, heeding the adage "try before you pry."
   b. Quickly check the exposure building by performing a quick 360-degree walk around all involved buildings to determine the easiest way to gain entry and limiting damage.
   c. Force entry into the exposure buildings immediately without stopping to consider the damage.
   d. Force entry into the exposure buildings immediately, but consider the damage.

4. The _____ axe head is not usually sharpened to as fine an edge as other axes.
   a. Flat   c. Pick
   b. Chisel   d. Adz

5. The rabbit tool is used to pry open doors and windows, but not
   a. Inward swinging doors
   b. Outward swinging doors
   c. Metal fire doors
   d. Metal-framed double-hung windows

6. It is usually easier to force entry through a
   a. Metal door
   b. Metal-clad wooden door
   c. Window

7. When breaking window glass from an upper story window, the fire fighter should
   a. Work from the upwind side and above the window
   b. Work from the upwind side and alongside the window
   c. Work from the downwind side and above the window
   d. Work from the downwind side and alongside the window

8. It is almost always easier to force entry through the _____ door of a store.
   a. Front
   b. Rear
   c. Side
   d. No statement can be made regarding which door is easier to force, as this depends on the type of store and location.

9. When confronting a commercial building with a metal door that is set in a masonry wall and the door must be forced open, the quickest method would be
   a. Placing the adz end of a pry tool in the doorframe, drive it in, and force it outward
   b. Using a rabbit tool
   c. Pulling the lock
   d. Removing exposed door hinges

10. A _____ lock is practically impossible to force.
    a. Rotating tumbler triple
    b. Straight double-bar bolt
    c. Detroit
    d. Fox

11. Sometimes it is necessary to breach a wall to enter an apartment. In this regard
    a. It is generally easier to breach the wall between the apartment and corridor.
    b. It is generally easier to breach the wall between apartments.
    c. Walls between apartments and corridors are of the same construction; thus, breaching either wall will require the same effort.
    d. Always breach the wall next to the door.

12. When a concrete block exterior wall is breached to gain entry
    a. Make the opening next to the door.
    b. Create an opening as near as possible to the center of the wall.
    c. Open the wall at a corner.
    d. Open the wall next to a corner.

Each of these presents special problems; therefore, none can be established as more or less difficult to force.

A complete teaching and learning system developed by educators with an intimate knowledge of the obstacles you face each day supports *Ladder Company Fireground Operations, Third Edition.* The resources provide practical, hands-on, time-saving tools like PowerPoint presentations, customizable lecture outlines, test banks and image/table banks to better support you and your students.

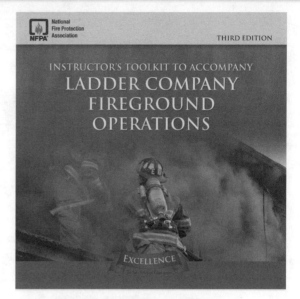

Instructor's ToolKit CD-ROM

ISBN-13: 978-0-7637-5235-4

ISBN-10: 0-7637-5235-5

Preparing for class is easy with the resources on this CD-ROM, including:

- **Adaptable PowerPoint Presentations**—Provides you with a powerful way to make presentations that are educational and engaging to your students. These slides can be modified and edited to meet your needs.
- **Lecture Outlines**—Provides you with complete, ready-to-use lecture outlines that include all of the topics covered in the text. Offered as word documents, the lecture outlines can be modified and customized to fit your course.
- **Electronic Test Bank**—Contains multiple-choice and scenario-based questions, and allows you to originate tailor-made classroom tests and quizzes quickly and easily by selecting, editing, organizing, and printing a test along with an answer key, including page references to the text.

The resources found on the *Instructor's ToolKit CD-ROM* have been formatted so that you can seamlessly integrate them onto the most popular course administration tools.

Acknowledgements

Reviewers

Scott Cullers
Loudon County Department of Fire-Rescue
Leesburg, Virginia

Ken Farmer
Former Fire Chief
Fuquay-Varina Fire Department
Fuquay-Varina, North Carolina

Todd Gilgren
Arvada Fire Protection District
Arvada, Colorado

Brian P. Kazmierzak, EFO
Clay Fire Territory
South Bend, Indiana

Ray McCormack
Fire Department, City of New York
New York, New York

Matt Szpindor
Fire Department, City of New York
New York, New York

Contributing Author

Stephen G. Persson
Captain
Cambridge Fire Department
Cambridge, Massachusetts

Captain Steve Persson is a thirty-five year veteran of the Cambridge Fire Department.

He was promoted to Captain in 1986 and assumed the role of Department Training Officer in 1991. He has retained this position for the past 16 years. The Training Division is responsible for the every day training of the 278 member department including basic and advance firefighting skills, emergency medical services, driver training, hazardous materials, and homeland security programs. Initial recruit instruction is also conducted after graduation from the Massachusetts Firefighting Academy.

Captain Persson has planned, coordinated, and lent support to several mass casualty exercises, hazardous materials, and weapons of mass destruction training programs in which the Cambridge Fire Department, mutual aid departments and the public and private sectors have participated. Captain Persson was instrumental in implementing an Incident Management System within the department that has been in effect for the past 14 years. Captain Persson holds an AS degree in Fire Science Technology, has attended the University of Massachusetts Institute of Government Services, is a certified Instructor I, and a state certified Harzardous Materials Technician. Captain Persson has been appointed by the Governor as a member of the Massachusetts State Fire Training Council and represents the interests of the Massachusetts Institute of Fire Department Instructors. He has worked at the National Fire Protection Association in Quincy, Massachusetts in the Public Fire Protection Division as a Fire Service Research Assistant, author, technical assistant, and consultant. In addition to his duties on the Cambridge, Massachusetts Fire Department, he is the Fire Chief and Town Fire Warden in the town of Frye Island, Maine.

Preface

Fire departments face many challenges. Ladder company personnel are an integral part of firefighting operations on the fireground. This book emphasizes that fire fighters performing ladder company tasks must be properly trained, possess the proper equipment, and be adequately staffed. *Ladder Company Fireground Operations* covers the basic objectives of ladder company work including the assignments of conducting a primary search, rescuing victims, forcing entry, and conducting proper ventilation techniques. This book also emphasizes other areas of importance such as applying pre-incident planning, using standard operating guidelines, and working with an Incident Management System.

Ladder and engine company operations are both critical to the success of a fireground operation, and these efforts must be coordinated in a safe and effective manner. Ladder companies are responsible for removing heat, smoke, and gases to allow greater visibility and permit engine company personnel to move rapidly and safely within a fire building. The examples used in this book do not include all the duties of a ladder company, but rather illustrate some important points about ladder company work.

Ladder Company Fireground Operations provides the reader with a concise look at the issues facing many fire departments. This book, along with the companion book, *Engine Company Fireground Operations*, aims to assist fire departments as they develop and conduct effective operations. These books provide overall basic fireground procedures required for effective firefighting activities.

Chapter 1 of *Ladder Company Fireground Operations* explains the typical duties of a ladder company, how their efforts impact fireground operations, and what procedures are most effective. Chapter 2 focuses on the process of initial assignments, including the positioning of apparatus and personnel, sizing up the fireground, and making key decisions. The issue of rescue, which is the first and highest priority at an incident, is discussed in Chapter 3.

Chapters 4 and 5 provide an overview of ventilation techniques and operations, which aim to make a fire building safer for both occupants and fire fighters as they carry out search and rescue operations. Chapter 6 covers the issue of checking for fire extension to limit the spread of fire, while Chapter 7 covers the topic of forcible entry, including the tools and techniques used in this type of activity. The use of aerial operations and ground ladders at the fireground are discussed in Chapters 8 and 9. These chapters also cover the dangers and the limitations of using such equipment.

Chapter 10 discusses property conservation techniques to mitigate the impact of a fire on a building and its occupants. Chapter 11 covers the use of elevated streams to assist in firefighting operations and also addresses some of the hazards that should be avoided. Chapter 12 discusses how fire fighters take control of utilities at the fireground and emphasizes the importance of pre-incident planning to determine what specific utility issues will need to be addressed at a location. The importance of overhaul activities, which are dangerous but necessary to ensure that all fire is extinguished and the area is safe, is discussed in Chapter 13.

This book provides a comprehensive overview of ladder company operations at the fireground. The book also identifies the tools, procedures, and techniques that can be used for enhancing fire department operations to make the operations run as efficiently as possible.

Captain Stephen Persson
Cambridge Fire Department
Cambridge, Massachusetts

Introduction

Chapter 1

Learning Objectives

- Identify the six basic fireground objectives of a ladder company.

- Understand how preincident planning assists in decision making on the fire ground.

- Define fire spread and understand how fire advances through a building.

- Evaluate the three ways in which fire travels: convection, radiation, and conduction.

- Be familiar with the term flashover and its potential impact on the fire ground.

- Be familiar with the term backdraft and its potential impact on the fire ground.

- Identify the 10 basic tasks usually assigned to ladder companies.

The duties of a ladder company on the fire ground are challenging as well as dangerous. Fire fighters performing ladder company tasks must be properly trained, possess the proper equipment, and must be adequately staffed. Members need to be educated so that they will make the proper decisions while working in a hostile environment. They must be proficient in the basic objectives of ladder company work, including the assignments of conducting a primary search, rescuing victims, forcing entry, and conducting proper ventilation techniques.

Fire fighters can gain valuable insight into a building's features by engaging in preincident planning activities and conducting building inspections. These actions provide details relating to building construction, layout, content, special hazards, and fire protection systems, which can be extremely helpful during sizeup and continued operations at the incident. Preincident planning should assist fire fighters in making decisions that will support the operation and ensure the safety of the personnel working at the incident.

Ladder company personnel must be well disciplined, follow department standard operating guidelines, and work with the incident management system. The safety of all fire fighters working on the fire ground must be the number one priority. Ladder and engine company operations must be coordinated in a safe and effective manner. Members must be accountable to command and follow direct orders. Freelancing should not be tolerated on the fire ground. Firefighting is a dangerous and stressful occupation, and fire fighters must be mindful of this fact at all times. Knowledge, skill, and judgment are needed to ensure proficiency in the performance of their assigned tasks and ensure the safety of all fire fighters working on the fire ground.

A number of fire service officials have stated that the overall efficiency of a fireground operation is determined by the performance of the ladder company. Others, in an analogy with the military, consider engine companies as the fire-fighter infantry and ladder companies as the engineers. These statements characterize the importance of ladder company operations and the role of the ladder company on the fire ground.

Ladder companies provide access to and exits from all parts of a fire building. Their crews also are responsible for removing heat, smoke, and gases to allow greater visibility and permit engine company personnel to move rapidly and safely within a fire building or exposure buildings **Figure 1-1**. These examples

Figure 1-1   Ladder companies assist engine companies by gaining entry, laddering the building, controlling the fire spread using ventilation techniques, and evacuating occupants.

Key Points

Ladder companies provide access to and exits from all parts of a fire building. Their crews also are responsible for removing heat, smoke, and gases to allow greater visibility and permit engine company personnel to move rapidly and safely within a fire building or exposure buildings.

do not by any means include all of the duties of a truck company, but they do illustrate two important points about ladder company work.

First, the tasks performed by a ladder company are required at every fire, regardless of who carries them out. In a fire department that includes one or more ladder companies, the ladder work is clearly the responsibility of those companies. When a department does not include ladder companies, arrangements are often made to have neighboring ladder companies respond to first alarms, but even if ladder companies are not available within a department or neighboring departments, the tasks assigned to a ladder company must still be performed. In such situations, it is important that these tasks are assigned to particular personnel; that these fire fighters are thoroughly familiar with the tools, skills, and operations of ladder company operations; and that they are trained and assigned as a team.

Second, ladder company operations may either accompany or precede engine operations. In many fire situations, such as forcible entry, attack line advancement could be delayed unless a ladder company crew initially made entry into the building. If ladder company operations are performed inefficiently, other fireground operations are adversely affected; therefore, sufficient personnel must be assigned to these tasks whether or not the crew is assigned to a separate ladder company. The proper tools and equipment must be available and the crew assigned proficient in

Key Points

The tasks performed by a ladder company are required at every fire, regardless of who carries them out.

Key Points

Ladder company operations may either accompany or precede engine operations. In many fire situations, such as forcible entry, attack line advancement could be delayed unless a ladder company crew initially made entry into the building.

their use. The crew must be fully trained in all ladder company fireground operations.

Fire fighters performing ladder company operations should then have the training, equipment, and adequate personnel to carry out the six basic objectives of a firefighting operation. Listed in the order in which they must be accomplished, these six objectives are as follows:

1. Rescue victims
2. Protect exposures
3. Confine the fire
4. Extinguish the fire
5. Provide property conservation
6. Overhaul the fire ground

Property conservation is sometimes considered to be a firefighting objective rather than an operation. Although property conservation does not contribute directly to control of a fire, it is very important in limiting the fire loss. In this book, property conservation is considered a ladder company task, although any fire fighter could be assigned to perform this important function. It will be discussed, along with other ladder company tasks, in the chapters that follow.

The list does not contain any of the "mechanical" movements, or evolutions, involved in fire fighter—such as raising ladders, advancing hose lines, or ventilating. These operations are performed to accomplish one or more of the six basic objectives. For instance, a ladder company crew might raise a ladder to allow a trapped person to climb down safely, but the objective is rescue, not the raising of the ladder. Firefighting objectives are the same at every fire; a particular fireground situation dictates what movements are required to accomplish the objectives.

All of these objectives could be carried out in an atmosphere of flame and smoke. Thus, it is essential that fire fighters understand the nature of fire and the factors that affect its spread, including building construction, type of occupancy, and types of fuel available to the fire.

Fire Spread

Oxygen, fuel, and heat are required to start and sustain the combustion process. These form the three sides of the familiar

Key Points

It is essential that fire fighters understand the nature of fire and the factors that affect its spread, including building construction, type of occupancy, and types of fuel available to the fire.

fire triangle. A more advanced concept of combustion includes a fourth element, a chain reaction phase, to form a fire tetrahedron. The chemistry of fire is not covered in this text, nor are the technical aspects of the support of combustion. Fire fighters are confronted with the problem after the fact. The discussions in this text are directed toward understanding how fire advances through a building and how it extends to exposures, as these are the characteristics that affect fire-fighter operations.

In structural fire situations, the fuel and oxygen required to sustain a fire are generally in plentiful supply. The fire usually starts out small and, if attacked early enough, could easily be confined to the vicinity of its origin. When the fire burns unchecked, heat production increases. As the original fuel sources are consumed, the fire travels to new fuel sources in uninvolved parts of the building and in exposures. There are three ways by which fire travels: convection, radiation, and conduction.

Convection

Convection is the travel of heat through the motion of heated matter—that is, through the motion of smoke, hot air, heated gases, and flying embers, as shown in **Figure 1-2**.

When confined, as within a structure, convected heat moves in predictable patterns. The fire produces gases that being lighter than air rise toward the top of the building. Heated air also rises, as does the smoke that combustion produced. As these heated combustion products rise, cool air takes their place; the cool air is heated in turn and then also rises to the highest point it can reach. As the hot air and gases rise away from the fire, they begin to cool; as they do, they drop down to be reheated and rise again. This is the convection cycle. Within a building, this cycle will first fill the upper parts and then work down toward the fire.

It is easy to see how this method of heat and fire transmission creates a need for rescue operations and for checks of fire spread in the building. In addition, convection is the main reason for ventilation activities in fire department operations.

Modern aerial apparatus and required equipment have evolved over the years because of the convection problem in

Figure 1-2   Convection carries hot air, smoke, gases, and embers upward through available vertical channels.

old brick, wood-joist buildings. Such structures, having open stairways, elevators, dumbwaiters, and other shafts, many formed by their interior construction, permit rapid vertical spread of fire by convection.

Older building codes first passed in major cities did not require fire-resistant construction, standpipes, or other protection unless the building was over 75 feet high. This was probably because at the time these laws were passed the 75-foot aerial ladder was in great use throughout the country.

Modern codes aimed at limiting fire spread by convection require that buildings of more than three stories be of fire-resistant construction. Fire-resistant construction is intended to confine the convection cycle to one floor or to a small area of a floor in any type of building.

When fire is prevented from spreading upward, convection carries the fire outward. The gases spread across the ceiling and down walls and travel into adjacent rooms or areas **Figure 1-3**. Eventually the area is saturated with superheated gases. The intense heat quickly brings all combustible materials up to ignition temperatures, and the result is usually a violent production of flame that travels extremely fast from room to room.

Figure 1-3 When vertical travel is blocked, convection carries hot air, smoke, gases, and embers horizontally.

Key Points

When fire is prevented from spreading upward, convection carries the fire outward. The gases spread across the ceiling and down walls and travel into adjacent rooms.

Radiation

Radiation is the travel of heat through space; no material substance is required. Pure heat travels away from the fire area in the same way as light—that is, in straight lines. It is unaffected by wind and, unless blocked, radiates evenly in all directions **Figure 1-4**. After the fire has built to sizable proportions, radiation is the greatest cause of exposure fires, spreading fire rapidly from structure to structure or through storage areas. Intense radiant

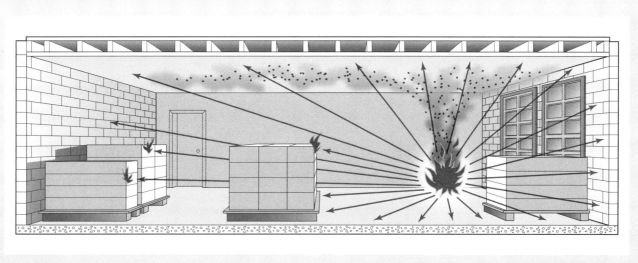

Figure 1-4 Heat is radiated evenly in all directions from the fire.

heat drives fire fighters back from normal approach distances, necessitating the use of master streams on the fire and exposed structures or stored materials.

In combination with convected heat, radiation creates the most severe area of exposure; this area must be protected first. However, although radiation is not affected by wind, the windward side of the fire cannot be ignored. Fire departments have been caught short too often when their efforts were directed solely to the area hit by both radiant and convected heat. After this most dangerous area has been covered, attention must be given to those areas exposed to radiant heat alone.

Within a building, radiant heat quickly raises the temperature of air and combustible material both near and, if the layout permits, at quite some distance from the fire. As a result, flashover can occur long before the flames actually contact the fuel in a given area. Flashover is addressed in more detail later in this chapter.

Proper ventilation is of little help against concentrations of radiant heat. Venting will remove the smoke, hot air, and heated gases, thereby lessening the chance of rapid spread and flashover, but the radiant heat remains and must be counteracted through proper application of water on the seat of the fire.

Radiant heat, by itself and in combination with convected heat, can cause great physical distress to the fire fighter. For this reason, it is imperative that full protective clothing be worn.

Conduction

Conduction is the travel of heat through a solid body **Figure 1-5** . Although normally the least of the problems at a fire, the chance of fire travel by conduction should not be overlooked. Conduction can take heat through walls and floors by way of pipes, metal girders, and joists and can cause heat to pass through solid masonry walls.

If the spread of fire by conduction occurs at all, the time involved will depend on the amount of heat and fire being applied to a structural member or wall. In any case, when fire has been in contact with such parts, a thorough check must be performed to ensure that the fire has not traveled through them to other areas. Fire fighters must also be aware that heat can be conducted down, as well as in other directions, depending on the building design and features in the fire area.

Conduction also can be dangerous to fire fighters. Certain types of structures have steel building and roof supports that

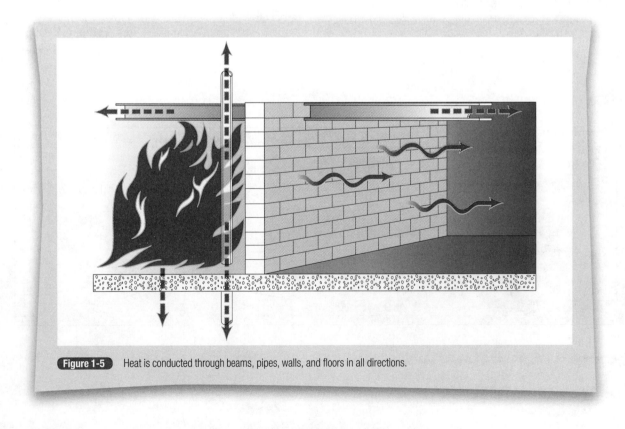

Figure 1-5   Heat is conducted through beams, pipes, walls, and floors in all directions.

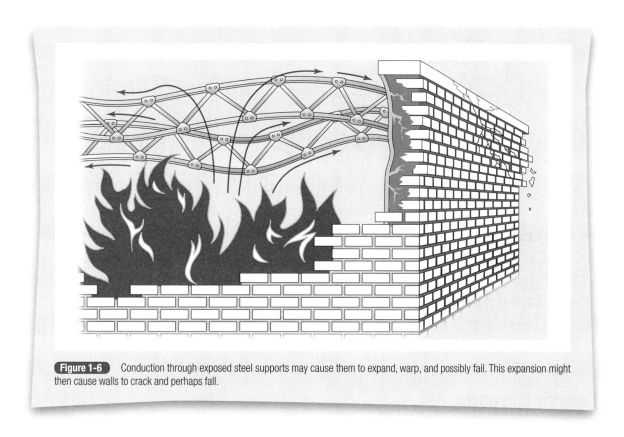

Figure 1-6 Conduction through exposed steel supports may cause them to expand, warp, and possibly fail. This expansion might then cause walls to crack and perhaps fall.

are completely open to fire. Heat spreading through these supports raises their temperature and can cause them to warp and fail, possibly causing the walls and roof to collapse **Figure 1-6** .

In many cases, hose streams stop conduction of heat by removing heat from the structural members, walls, and floors. Because heat traveling by conduction cannot be seen or felt in the normal course of fire-fighter operations, the fire fighter must be observant and check for the spread of fire by conduction when there are indications that this could take place.

Flashover

Flashover is the ignition of combustibles in an area heated by convection, radiation, or a combination of the two. The action can be a sudden ignition in a particular location, followed by rapid spread, or a "flash," of the entire area. The latter action is more likely in an open area within a building.

Convection can cause flashover at the top of a structure because of the hot products of combustion igniting materials at that level. Radiation can contribute to flashover in areas that do not block heat travel. Flashover, however, is not usually caused by radiation alone; radiation in combination with convection is more often a cause of flashover **Figure 1-7** .

Smoldering Fire and Backdraft

The products of a fire can fill a building until the fire is almost starved for oxygen, at which point it will begin to smolder. The more incomplete the combustion, the more carbon monoxide is produced. Carbon monoxide is a colorless, odorless, poisonous gas, especially dangerous because it is also explosive and flammable. Two of the elements necessary to produce fire—heat and fuel—are contained within the structure: Only oxygen need be added. If oxygen is improperly allowed to enter the structure, the accumulated gases will ignite into a rapidly spreading fire or a violent explosion. This is **backdraft**. Fortunately, this situation can be controlled effectively through proper ventilation and attack procedures.

Key Points

Flashover is the ignition of combustibles in an area heated by convection, radiation, or a combination of the two. The action can be a sudden ignition in a particular location, followed by rapid spread, or a "flash," of the entire area.

Figure 1-7 Flashover, heated gases, and radiant heat traveling through a building may cause ignition at some distance from the fire.

Ladder Company Operations

Ladder companies are sometimes referred to as "truck" companies, "hook and-ladder" companies, "aerial ladder" companies, elevating platforms, "tower ladder" companies, and "snorkel" companies. Such labels might partially describe the particular type of apparatus used by ladder companies, but they do not indicate the personnel, equipment, and the amount of planning and training that is needed to operate a ladder company efficiently and safely.

Ladder company apparatus and equipment have been designed to permit crews to function effectively and quickly in accomplishing the six fire-fighter objectives. Through training and experience, ladder company personnel acquire knowledge, skill, and judgment in performing the basic tasks usually assigned to ladder companies. These tasks include but are not limited to the following operations:

- Conduct the primary search
- Rescue trapped victims
- Ventilate the building
- Ladder the building
- Force entry
- Check for fire extension
- Provide property conservation
- Control utilities
- Overhaul
- Salvage
- Operate elevated master streams

At some fires, it might be necessary for a ladder company to perform all of these operations; other fires might require only some of the tasks to be carried out. Just as situations vary, procedures for each situation will also vary. With the exception of rescue, the tasks may not necessarily be performed in the order given previously here; that, too, depends on the fire situation.

For example, at a vacant building heavily involved in fire, the first ladder company task might be to operate an elevated master stream. Other ladder company tasks at such a fire might be limited to checking for fire extension in adjoining or nearby buildings, laddering, and forcing entrance into these structures. On the other hand, a working fire in an occupied building could easily require that every ladder company task be performed.

Knowledge of the company's district, building inspections, and preincident planning are necessary if ladder companies are to operate efficiently on the fire ground **Figure 1-8**. This information is also useful in developing or improving on standard

Key Points

Knowledge of the company's district, building inspections, and preincident planning are necessary if ladder companies are to operate efficiently on the fire ground.

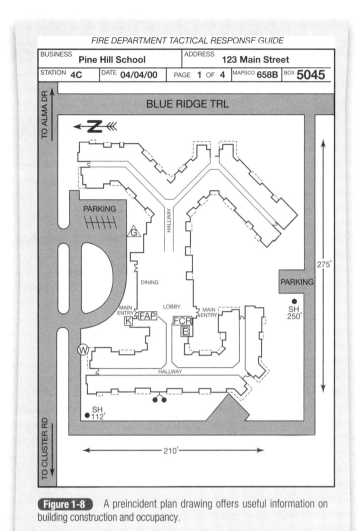

FIRE DEPARTMENT TACTICAL RESPONSE GUIDE

BUSINESS **Pine Hill School**	ADDRESS **123 Main Street**		
STATION **4C**	DATE **04/04/00**	PAGE **1** OF **4**	MAPSCO **658B** BOX **5045**

TO ALMA DR

BLUE RIDGE TRL

N

PARKING

HALLWAY

G

DINING

PARKING

275'

LOBBY

MAIN ENTRY

MAIN ENTRY

SH 250'

K FAP

FCR E

W

HALLWAY

SH 112'

210'

TO CLUSTER RD

Figure 1-8 A preincident plan drawing offers useful information on building construction and occupancy.

operating guidelines. Ladder company personnel should have a good knowledge of their first alarm district, especially with regard to building construction and occupancy, life hazards, exposure hazards, and locations especially dangerous to fire fighters. Other special conditions that affect fire fighter procedures such as access to the property, laddering, and ventilations concerns should be addressed by the members. Perhaps it is impossible to learn everything about the company's district; however, unusual situations and target hazards should be carefully examined and analyzed, and special procedures should be developed when necessary.

This background information can be of particular help when the company arrives at a working fire and begins initial sizeup. A continuing procedure, sizeup results in operational changes to match changes in the fire situation. Factors to be considered in sizing up a fire situation and the ladder company operations required by different fire situations are covered in the remainder of this book.

Wrap-Up

Chief Concepts

- There are 10 basic ladder company tasks. Each of these tasks may or may not be required during a particular fire situation.
- Ladder companies should be staffed, equipped, and trained to perform necessary duties quickly and efficiently, as their performance can affect the safety and performance of other responding companies.
- Ladder company operations should take into account the ways in which fire, heat, and smoke travel through a building, as well as the knowledge gained by preincident planning and company inspections.
- Ladder company work can be one of the most challenging and rewarding assignments, provided that the ladder companies in a department operate properly and safely.
- Ladder companies must not be considered solely for the use of their aerial device, or worse, only for elevated master streams.
- The efficiency of ladder company operations is a most important factor in determining the overall fireground efficiency.

Key Terms

Backdraft: When oxygen is improperly allowed to enter the structure rapidly, the accumulated gases will ignite into a rapidly spreading fire or a violent explosion.

Conduction: The travel of heat through a solid body. Although normally the least of the problems at a fire, the chance of fire travel by conduction should not be overlooked. Conduction can take heat through walls and floors by way of pipes, metal girders, and joists and can cause heat to pass through solid masonry walls.

Convection: The travel of heat through the motion of heated matter—that is, through the motion of smoke, hot air, heated gases, and flying embers.

Flashover: The ignition of combustibles in an area heated by convection, radiation, or a combination of the two. The action can be a sudden ignition in a particular location, followed by rapid spread, or a "flash" of the entire area.

Property conservation: Procedures and operations used to protect and reduce damage to the building and its contents from the effects of smoke, heat, fire, and water damage during firefighting operations.

Radiation: The travel of heat through space; no material substance is required. Pure heat travels away from the fire area in the same way as light—that is, in straight lines. It is unaffected by wind and, unless blocked, radiates evenly in all directions.

1. Ladder company tasks must
   a. Always be performed by an established ladder company from the jurisdiction where the fire occurs
   b. Always be performed by a ladder company from the jurisdiction where the fire occurs or by a mutual aid ladder company
   c. Always be performed, regardless of who carries them out
   d. None of the above

2. In structural fire situations, oxygen, fuel, and heat are needed to sustain combustion. Of these, two are generally in plentiful supply. Which of the following is normally the missing component that is needed to initiate and continue fire spread?
   a. Fuel
   b. Heat
   c. Oxygen
   d. Actually, the chemical reaction is most critical.

3. What are the ways by which fire travels?
   a. Convection
   b. Radiation
   c. Conduction
   d. All of the above

4. Convection is the travel of heat
   a. Through the motion of heated matter
   b. Via chemical reaction
   c. Through space
   d. Through a solid body

5. Radiation is the travel of heat
   a. Through the motion of heated matter
   b. Via chemical reaction
   c. Through space
   d. Through a solid body

6. Conduction is the travel of heat
   a. Through the motion of heated matter
   b. Via chemical reaction
   c. Through space
   d. Through a solid body

7. In a structure fire, ventilation is most effective in reducing _____ heat.
   a. Convected
   b. Conducted
   c. Radiant
   d. Ventilation is equally effective on all of these heat types.

8. Flashover is the ignition of combustibles that are heated by
   a. Convected heat
   b. Conducted heat
   c. Radiant heat
   d. Either convected, radiant heat, or a combination of radiant and convected heat

Initial Assignments

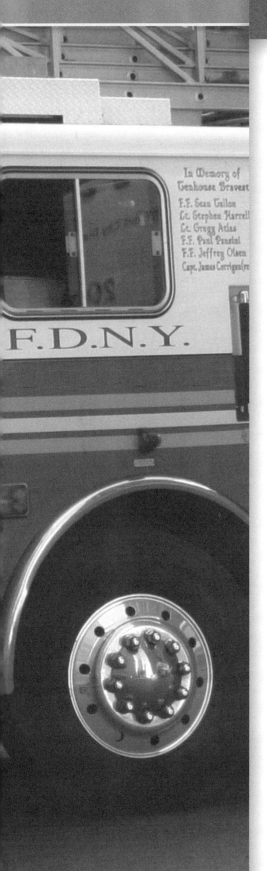

Chapter 2

Learning Objectives

- Identify the equipment carried on a ladder truck, including ground ladders, hand and power tools, and other associated equipment.

- Assess the differences between hand tools and power tools and how they are used on the fire ground.

- Understand that ladder company personnel should be assigned to particular tools based on riding positions and/or designated tasks.

- Recognize that a ladder company should be capable of operating efficiently from the moment it arrives on the fire ground.

- Realize the importance of an initial sizeup that will assist in critical decision making during the initial stages of the fire.

- Understand that ladder company apparatus must be positioned properly on the fire ground to accomplish the particular task required.

To do their job, fire fighters must arrive safely on the fire ground. Fire apparatus must be driven with caution and with due regard for the safety of both the general public and the fire fighters riding aboard the apparatus. On arrival at the fire ground, a sizeup should be made to determine the current conditions and whether additional resources will be needed to perform all ladder company functions.

Fire apparatus should be positioned in a location that will affect the particular task or tasks required, including the operation of the aerial device. The protection of fire fighters and the apparatus should be considered when approaching the incident. Many variables need to be considered when safely positioning the apparatus. Included in the list are a check for overhead wires, branches, and other obstructions, ensuring that the apparatus is on stable ground and that a safe operating angle has been achieved to perform assigned tasks. After the apparatus is positioned, it may be difficult to reposition or relocate. Operators should take into considerations the need for access and departure by other emergency vehicles such as tankers and ambulances.

Fire fighters must be in proper personal protective equipment and self-contained breathing apparatus when they arrive on scene. Members should be assigned specific functions and carry the proper tools and equipment needed to perform their tasks safely.

Priority must be given to the primary search and rescue of victims. Adequate ladder company personnel must be on scene to perform the required tasks of rescue safely and efficiently. Preincident planning and inspections of buildings can assist fire fighters in making appropriate decisions because they have a working knowledge of the property. This should assist them in making decisions that will support the operation and ensure the safety of the fire fighters working at the incident.

Most fire departments—paid or volunteer—operate with a limited number of personnel. If a ladder company is to be ready for any fireground situation, it must be configured so that assigned personnel can perform all of the required tasks in as short a time as possible. Response patterns and operating procedures must ensure that members get into action quickly and that the most urgent operations are begun immediately after arrival on the fire ground.

For this, each member must be trained in and thoroughly familiar with the use of all of the tools and equipment on the apparatus. Before the ladder company arrives on the fire ground, all crew members should know which tools they will be carrying into the fire building and what their duties will be on arrival—that is, their initial assignments. On arrival, command should assign which areas of a fire building each ladder company is to cover. Standard operating guidelines will dictate how the apparatus is to be positioned and, based on command's assignment of that company, what duties they are to perform given the particular fire

Key Points

Both the company officer and each fire fighter assigned to that ladder company must know how to perform their initial assignments when they arrive on the fire ground.

situation. Both the company officer and each fire fighter assigned to that ladder company must know how to perform their initial assignments when they arrive on the fire ground.

This chapter deals with the tools and equipment carried on the truck, the assigning of responsibility for these tools to the members, coverage of the fire building, and apparatus positioning. Included in the discussions are some simple but effective operating procedures to ensure that all ladder company fire fighters are aware of their initial assignments and able to accomplish them; however, it is important to realize that all firefighting procedures require training if they are to be performed efficiently in a fire situation.

Tools and Personnel

A ladder company is designed to carry ground ladders, associated tools and equipment, and typically an aerial device. The sizes and types of ladders carried on a ladder will depend on minimum standard recommendations and department experience. Fire fighters must be trained to carry, raise, and climb ground ladders and must be ready to place them properly when they are required on the fire ground. The apparatus should be positioned so that the aerial device can be placed into operation if needed. Aerial operations and ground ladder operations are discussed in detail in Chapters 8 and 9. Tools are the most used items on a ladder truck. Because all ladder company tasks require tools, they are used at almost all incidents.

Hand Tools

Hand tools are usually categorized as cutting, prying, or push/pull devices. Hand tools most useful for forcible entry include the claw tool, crow bar, the Kelly tool, the Halligan bar, Quic Bar, and similarly designed tools. Other tools used for this task include lock pullers such as the K tool, bolt cutters, and the flathead ax and pick-head ax. These tools can be used to break locks, to force exterior and interior doors, to open sidewalk doors and grates, and generally to provide entry into a building.

Key Points

Hand tools most useful for forcible entry include the claw tool, crow bar, the Kelly tool, the Halligan bar, Quic Bar, and similarly designed tools.

Key Points

For ventilation work, the pick-head ax, pike pole, plaster hook, and Halligan bar or similarly designed tools are usually preferred.

The flathead ax also can be used to drive other tools, including another flathead ax.

For ventilation work, the pick-head ax, pike pole, plaster hook, and Halligan bar or similarly designed tools are usually preferred. The ax can be used to cut and rip open roofs, to force or knock out windows, and to cut and force natural or built-in openings such as skylights, ventilators, and roof hatches or scuttles. The Halligan bar is used to rip up cut roofing, to force and knock out windows, and to force natural openings.

The pike pole (ceiling hook) can be used in ventilation work to pull down a ceiling or to push the ceiling down from the roof after the roof has been opened. It is also useful for opening a transom below a skylight, removing coverings below roof hatches, and knocking out window glass.

All of these tools can also be used in checking for fire extension and in overhaul. For these operations, they are used to pull down ceilings and walls, rip up floors and baseboards, and generally open the interior of a building for inspection for fire spread and extinguishment.

Power Tools

Power tools can be of great help in ladder company operations, especially where a lack of adequate personnel is a problem. Electricity, gasoline engines, or air or hydraulic pressure can power the tools. Power saws are used for cutting and include circular, chain, and reciprocating varieties. Powered hydraulic spreaders and cutters are other tools used to ease and speed up the work of the ladder company. Power tools have one major disadvantage: They are larger and heavier than hand tools and can take more time to get into position. They will operate much faster than hand tools, however. Large power tools are also more difficult to operate in confined spaces or areas where the tool just does not fit. In this case, hand tools may be the better choice. Power tools cannot replace hand tools for every application, and thus, a combination of both should be given careful consideration when equipment is being purchased.

Key Points

Power tools can be of great help in ladder company operations, especially where a lack of adequate personnel is a problem.

Training

Ladder crews must be thoroughly familiar with the tools of their trade. This means that they must be trained in the use of both hand and power tools, as well as in the duties of carrying, raising, and climbing ground ladders and operating the aerial device. Most ladder companies can find places in their response area where they can raise ground ladders, position the aerial device for rescue or ladder-pipe operations, and carry out other training activities. The department's training facility, the fire station, or target hazards may be places where members could practice their skills; however, ladder company tools and associated equipment are used to break things, and thus, training usually involves the destruction of property. Opportunities for complete training are thus quite rare, as no building owners want the doors of their buildings forced open, their roofs cut up, or their windows knocked out simply for training purposes.

Because the weakest part of the continuing training program will be tool use, fire fighters should take advantage of every training opportunity that becomes available. Many fire departments or training academies may be able to conduct their own training sessions using both hand and power tools. Buildings that are about to be torn down can, with permission, be used as tool-use training structures. Companies might also be allowed to use their tools in parts of a building about to be remodeled. In either case, proper supervision and strict safety standards must be in place. Mockups of buildings also have training use. The best training structure is one in which ladder crews can employ all of their tools as they would in an actual fire situation, and thus, they are prepared to perform all necessary fireground operations.

Tool Assignments

The ladder officer should not have to tell each member of the crew which tools to take into the fire or exposure building.

Key Points

Ladder crews must be thoroughly familiar with the tools of their trade. This means that they must be trained in the use of both hand and power tools as well as in the duties of carrying, raising, and climbing ground ladders and operating the aerial device.

Key Points

The ladder officer should not have to tell each member of the crew which tools to take into the fire or exposure building. Instead, as a part of preincident planning, personnel should be assigned to particular tools based on the locations of the tools on the truck and their riding positions.

Instead, as a part of preincident planning, personnel should be assigned to particular tools based on the locations of the tools on the truck and their riding positions.

Hand tools and power tools should all be located on the truck so that they are readily available. Such hand tools as axes, claw tools, Halligan bars, pike poles, lock pullers, and Kelly tools should be located at or near the positions in which the crew will ride to the fire. Assignment of a member to a particular truck position then should include responsibility for the tools near that position. Fire fighters should take these tools into the building, either immediately after arrival or after performing necessary outside duties.

The ladder company officer must ensure that the tools are properly located on the truck and must assign members to riding positions and tools in such a way that the proper assortment is taken into the building.

In paid departments, fire fighters should be assigned riding positions and tools when their shift comes on duty. In some departments, all members are assigned positions on the truck, and this determines the tools for which they are responsible. In other departments, all members are assigned groups of tools, and this determines the positions that they will take on the truck. Either system is effective if followed consistently.

In some municipal departments, one or more members of a ladder company are assigned to forcible entry tasks and works with engine company personnel after arrival at the fire. This fire fighter is equipped with a claw tool, a Halligan tool or similar device, and a flathead ax and might also be equipped with a hydraulic spreading device like the rabbit tool. **Figure 2-1** shows a fire fighter equipped with some common tools for these tasks. This member forces entrance for the engine crew, ventilates the fire floor around them, and searches for victims in the fire area.

The person responsible for forcible entry tasks is assigned to a ladder company, rather than an engine company. If assigned to an engine company, this fire fighter soon would be advancing hose lines and not performing the required ladder company tasks.

In volunteer departments, tool assignments should be based on the positions taken by members on the apparatus. A chart of the truck showing crew positions and tool locations can be helpful in making these assignments. The chart should be kept on display in the station and made a part of the training program **Figure 2-2** . It may include a member in charge of forcible-entry tools to be assigned to work with engine personnel, as noted previously.

Volunteers who are in the station before an incident should be assigned a riding position, according to the chart. Priority positions should be assigned first. This immediate assignment eliminates the confusion that might result if several members attempted to take the same position when an alarm is received.

Figure 2-1 A fire fighter assigned to forcible entry tasks should be well equipped with the necessary tools.

Volunteers who arrive at the fire after the truck is positioned should report to the command post for assignment and then check to see that the necessary tools have been removed and taken into the building.

Consideration must be given to the fact that every incident will not require the same tool. Fire fighters assigned to specific tools should know, through training and experience, which implements—hand or power tools—will work best for that particular incident.

Coverage Response

A properly equipped, adequately staffed, and well-trained ladder company should be capable of operating efficiently from the moment it arrives on the fire ground. For this to occur, it is necessary to develop a standard response procedure to get the apparatus to the scene safely and in a minimum amount of time.

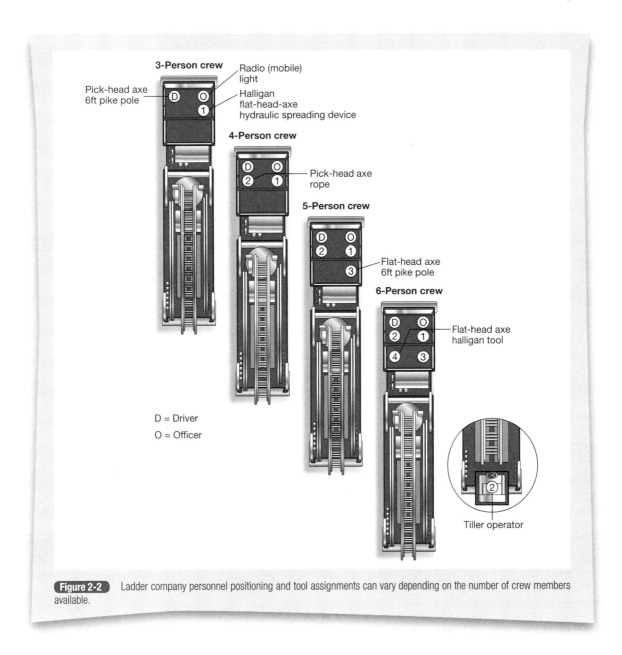

Figure 2-2 Ladder company personnel positioning and tool assignments can vary depending on the number of crew members available.

The details of any standard response procedure will vary according to the types of structures, their construction features, and the occupancy of the fire and exposure buildings. The procedure will also be affected by the number of companies responding to a first alarm, the distance between their stations, and the normal personnel complement of each responding company. Other considerations include the time of day and traffic, road, and weather conditions. The apparatus may be in the station or on the road, possibly out of their response area.

On arrival at an incident, a fire department must operate using a structured **incident management system (IMS)** on the fire ground. The IMS is an organized system of roles, responsibilities, and standard operating guidelines used to manage emergency operations. Ladder companies must work within an overall strategic plan developed by the **incident commander** **Figure 2-3**. Each ladder company must work within the system and follow the plan by performing tasks assigned by the incident commander ensuring that fire-fighter safety is maintained. The days of freelancing on the fire ground are over, and all fire fighters working at an incident must be accountable for their actions. Standard operating guidelines must be in place, and ladder

Key Points

A properly equipped, adequately staffed, and well-trained ladder company should be capable of operating efficiently from the moment it arrives on the fire ground. For this to occur, it is necessary to develop a standard response procedure to get the apparatus to the scene safely and in a minimum amount of time.

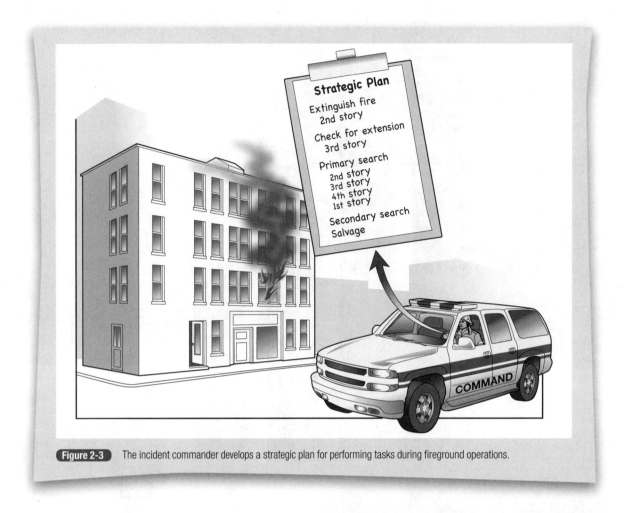

Figure 2-3 The incident commander develops a strategic plan for performing tasks during fireground operations.

companies should adhere to these procedures during fireground operations. The four primary tasks of a ladder company on arrival at an interior (offensive) operation are conducting the primary search, rescuing victims, ventilation, and forcible entry.

Front and Rear Coverage

To be effective, a standard response procedure must provide for immediate front and rear coverage of the fire building by the initial first-alarm companies. The sides of the building should then be checked to determine whether any immediate action needs to be addressed in these areas. Fire conditions often vary between the front and the back or sides of a building, as shown in **Figure 2-4**. It is important that the entire building be checked as soon as possible after arrival.

Key Points

Ladder companies must work within an overall strategic plan developed by the incident commander. Each ladder company must work within the system and follow the plan by performing tasks assigned by the incident commander ensuring that fire fighter safety is maintained.

Key Points

To be effective, a standard response procedure must provide for immediate front and rear coverage of the fire building by the initial first alarm companies.

If two ladder companies are responding to an alarm, the ladder company that is expected to arrive first (the first due) should be assigned to cover the front of the building. The second ladder company is assigned to the rear. These assignments should be modified according to the situation found on arrival. If the first due company finds a life hazard or some other serious situation

Key Points

If two ladder companies are responding to an alarm, the ladder company that is expected to arrive first should be assigned to cover the front of the building. The second ladder company is assigned to the rear. These assignments should be modified according to the situation found on arrival.

Figure 2-4 Fire conditions vary from the front (*left*) and to the rear (*right*) of an involved structure.

in the rear, it should cover that position first; the second due company should be notified by radio of the change in procedure and advised by the incident commander as to where to set up on the fire ground.

The assignment of a ladder company to rear coverage does not mean that the apparatus itself must be driven to the rear. In some cases, this is impossible. It does mean that company members must check the rear to determine what the situation is with regard to possible victims and the extent and intensity of the fire. They should check the locations of exposures and of rear stairs, porches, and basement entrances that might be used in firefighting operations. They should determine whether assistance is required by engine company personnel working

Key Points

If only one ladder company responds to the first alarm, at least one person from the company should be assigned to check out the rear immediately after arrival.

in the rear. A check on the availability of windows should be made to determine whether they could be used for ventilation, removing victims, or advancing attack lines. If necessary, ladder crews should provide ground ladders to upper stories. If only one ladder company responds to the first alarm, at least one person from the company should be assigned to check out the rear immediately after arrival.

To reach the rear of a fire building, ladder company personnel might have to move through adjoining buildings, through the lower levels of the fire building, or over row buildings, as shown in **Figure 2-5**. Any alleys, walkways, courts, and backyards can be used to gain access to the rear of a building. In some cases, it may be best to position the apparatus on the street at the rear of the fire building for both rear coverage and ladder company operations. Positioning of the apparatus has much to do with the efficiency of the coverage area when using the aerial device for rescue, gaining access, and using master stream appliances. Ladder company members will gain the proper skills to operate efficiently and safely through training, experience, preincident planning, and inspections.

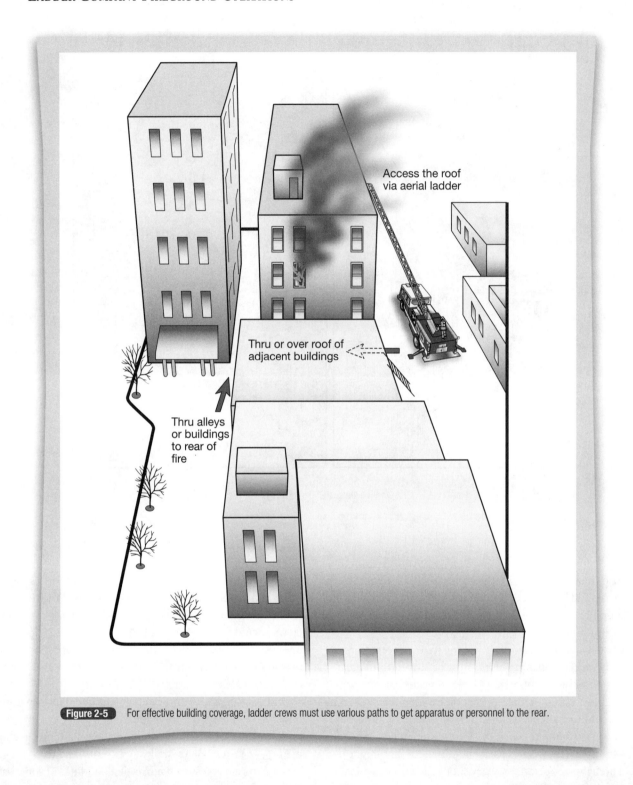

Access the roof
via aerial ladder

Thru or over roof of
adjacent buildings

Thru alleys
or buildings
to rear of
fire

Figure 2-5   For effective building coverage, ladder crews must use various paths to get apparatus or personnel to the rear.

Ladder Company Out of Quarters

Standard operating guidelines involving response procedures must
assume that companies will respond from quarters. Occasionally, a
ladder company will receive an alarm by radio when it is completely
out of position, perhaps in one corner of its district. The company
should advise the dispatcher of the situation by radio as soon as
the alarm is received. The dispatcher can then either send a closer

Key Points

The standard response procedure must apply to responding engine
companies as well as to ladder companies.

ladder company from quarters or advise the second-due ladder truck that it will be the first to arrive.

Other Aspects

Ladder companies do not work alone at a fire. The standard response procedure must apply to responding engine companies as well as to ladder companies. Reconnaissance of a building is the responsibility of both the engine and ladder companies and must be coordinated so that the companies responding to a first alarm work as a team with minimal duplication of effort. This is the job of the incident commander, whether it is the first-arriving company officer or the chief of the department. The important point is that the fire building must be covered quickly, front and rear, outside and inside, in a coordinated effort, as shown in **Figure 2-6**. Fire departments must have a written risk versus benefits analysis standard operating guideline, and it must be followed. For fire-fighter safety, vacant or abandoned buildings should not be entered if a major fire is encountered. In addition, any other situation that may jeopardize the safety of the fire fighter should be considered by the incident commander.

Sizeup Information

On arrival at the incident, an initial sizeup should be conducted by the first arriving crew. The building type and occupancy are

Figure 2-6 A standard operating guideline should aim to cover the fire building from the front and rear and both outside and inside the building in a well-coordinated effort.

critical factors in sizeup. The day of the week, the time of the day, and the weather conditions are also important considerations. In addition, reconnaissance of the building, observing what is taking place on the fire ground and communicating with other fire fighters, occupants, and bystanders, will assist the incident commander in the critical decisions made during the initial phase of operation. Company or sector officers should report pertinent information promptly to the incident commander. Whenever possible, reporting by radio will speed up the delivery of information. The information should be precise and accurate. Negative situations—such as when a position cannot be reached, when a task cannot be accomplished, or when requested information cannot be obtained—also should be reported. With reports from all areas, command should quickly develop a good mental picture of the situation. The fire fighter's sizeup will determine which operations should be initiated next and whether to call for additional companies.

Both the fire situation and the number of ladder company personnel at the scene affect the decision on which operations to initiate. First priority must be given to operations whose objective is rescue, conducting the primary search and removing victims; second, to exposure protection; third, to confining and extinguishing the fire; and so on through the list of objectives in Chapter 1, in the given order. Property conservation operations should be performed where possible to limit water damage. The smaller the crew, the more important it is that duties be assigned according to these priorities.

The number of personnel available determines how much can be done safely and effectively at one time. Five or six fire fighters might simultaneously conduct the primary search, force entry, and ventilate. A three-member crew might find it difficult to move through an area trying to perform three different tasks, all of which were needed to accomplish their objective, rescue, in a timely manner. How these operations will be performed is determined by sizeup. Although it is most desirable to enter the building and search for victims, the building might have to be

Key Points

Because ladder company operations affect the entire firefighting effort, command should not hesitate to call in additional companies as needed. It is extremely important that all necessary ladder company objectives be completed as quickly as possible.

Key Points

If in certain areas of a fire department's district more than one ladder crew will obviously be required, then additional ladder companies should respond to first alarms in those areas.

ventilated before it can be entered. In some cases, ladders will have to be raised to ventilate the building, or forcible entry may be required so personnel can get in to search.

Because ladder company operations affect the entire firefighting effort, command should not hesitate to call in additional companies as needed. It is extremely important that all necessary ladder company objectives be completed as quickly as possible. These operations should be performed simultaneously.

If in certain areas of a fire department's district more than one ladder crew will obviously be required, then additional ladder companies should respond to first alarms in those areas. The additional companies can be part of the department in whose district the fire occurs, made up of engine company crews assigned ladder company tasks, or they can come from neighboring departments. The important thing is that they are on their way to the fire at the first alarm or at any other time they are needed. If sizeup indicates that they will not be needed, the incident commander, can let them know by radio, and they can return to service; if they are needed, they will arrive on the fire ground in time to help save lives and avert undue property losses.

Apparatus Positioning

For the most effective operation, ladder company apparatus must be positioned properly on the fire ground to accomplish the particular task or tasks required.

Key Points

For the most effective operation, ladder company apparatus must be positioned properly on the fire ground to accomplish the particular task or tasks required.

Key Points

Standard operating guidelines and preincident plans should not require that the ladder truck be positioned in a certain way at every fire or that ladder crews perform certain duties regardless of fire conditions.

Standard operating guidelines and preincident plans should not require that the ladder truck be positioned in a certain way at every fire or that ladder crews perform certain duties regardless of fire conditions. The ladder truck need not be positioned if the aerial device is not needed. Such rules, including stipulation that the aerial should be raised at every fire for training purposes, are questionable. They lead to poor positioning, unnecessary operations, and wasted time. If a company is operating with inadequate personnel, each fire fighter must be used to perform the necessary tasks, not outside raising an aerial that is not going to be used. Training should be done at training sessions. Ladder companies must be positioned and assigned as each fire situation dictates.

Approach

Ladder trucks that are not committed to a specific task on the initial alarm should be placed in a staging area away from the fire building. When the incident commander assigns a ladder company to a task, the apparatus can leave the staging area and report directly to the assignment. In this way, the ladder company is given direct orders and clearly knows what its task is. This eliminates freelancing and enables command to manage the span of control. After a company is positioned on the fire ground, it may be difficult to move it to another location. Ladder companies responding on additional alarms should also be assigned to a staging area unless directed otherwise by command.

A ladder truck is not the easiest vehicle to maneuver in tight quarters or in close proximity to other apparatuses. The closer the apparatus is positioned in relationship to the fire building, the more difficult it is to move. The approach to the fire building therefore is very important.

After the ladder truck is at the fire scene, it should be advanced in a slow, deliberate approach to the fire building. No company should try to beat another company to a position; such foolishness can only lead to poor positioning, accidents, and injuries. The

Key Points

Ladder trucks that are not committed to a specific task on the initial alarm should be placed in a staging area away from the fire building. When the incident commander assigns a ladder company to a task, the apparatus can leave the staging area and report directly to the assignment.

Key Points

After the ladder truck is at the fire scene, it should be advanced in a slow, deliberate approach to the fire building.

Key Points

It is imperative that the ladder truck be positioned for a safe operating angle and has the reach to perform assigned tasks. The truck should be repositioned if it interferes with proper fireground operations.

officer in charge should be concerned only with getting a good position from which the company can work efficiently. The officer's approach to this position must be made in accordance with standard operating guidelines and the fire situation.

Positioning

Before a ladder truck is committed to a position of operation on the fire ground, fire fighters should observe some simple rules.

- Be aware of the positioning needs of other fire apparatus, both engine and ladder trucks, at all times.
- Allow room for engine companies to work in.
- Do not block hydrants or standpipe and/or sprinkler connections.
- Do not compromise egress to or from a building.
- Check the overhead for wires, tree branches, or other obstructions when final positioning is achieved.
- Ensure that the area under the ladder truck is stable and will carry the load.
- Check to see whether any tires of the apparatus are parked on supply or attack hose lines.
- Position the apparatus so that ground jacks will fully extend.

It is imperative that the truck be positioned for a safe operating angle and has the reach to perform assigned tasks. The truck should be repositioned if any of these items interferes with proper fireground operations.

Ladder companies should be prepared to relocate during fireground operations. Trying to leave an open area so that this may be accomplished may be difficult and sometimes impossible because of the amount of apparatus on scene. If conditions deteriorate, it may be necessary to move apparatus away from the fire building or exposures.

In areas of one- and two-story buildings, it is not usually essential that the ladder truck be positioned directly at the front (or rear) of the fire building. Because the aerial device probably will not be used in such low-rise structures, the truck can be positioned to one side of the building. This will leave the priority positions for the engine companies but will not hinder ladder company operations. Ladder crews will be able to remove ground ladders and tools and equipment from the truck quickly as required.

If the fire building is more than two stories high, the engine companies must allow the ladder companies to get close to the building so that aerial devices can be positioned effectively and ground ladders can be raised quickly. For example, suppose an engine company and a ladder company are approaching the front or rear of a fire building from the same direction, with the pumper ahead of the ladder truck. The engine company should pull on past the building to allow the ladder truck to be positioned properly, as shown in **Figure 2-7**. If the sides of the building are detached, this will also allow the engine officer to

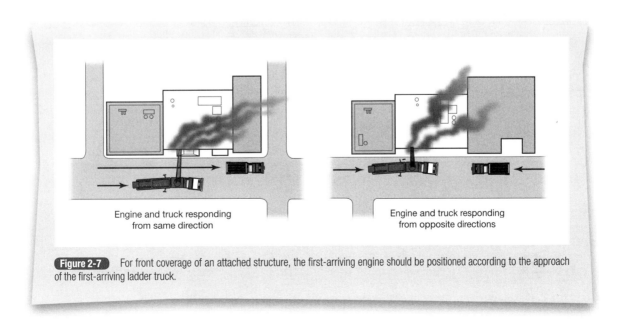

Engine and truck responding
from same direction

Engine and truck responding
from opposite directions

Figure 2-7   For front coverage of an attached structure, the first-arriving engine should be positioned according to the approach of the first-arriving ladder truck.

see three sides of the structure. The engine officer should report conditions on the far side to the ladder officer, who will have seen only two sides.

If an engine company and a ladder company are approaching a taller building from opposite sides, the pumper should stop short of the building. This, again, will allow the ladder company to get its apparatus into a good position. In this case, each officer will have seen two sides of the building; they should exchange information quickly, as operations begin.

In some cases, the width and or length of the building and the location of the fire within the building may solve the problem of positioning the ladder trucks and pumpers. At a fire in a large apartment or commercial building, the engine would be positioned near an entrance to allow engine crews to advance into the building and control the fire in stairways and corridors.

The ladder truck should be positioned for rescue, ventilation, or other ladder company operations **Figure 2-8** .

On the other hand, the extent, location, and intensity of the fire might not allow for perfect positioning of the ladder truck. In these cases, the truck should be positioned in a convenient and safe area and operations initiated without delay. If the building height and the intensity of the fire require such action, the apparatus should be positioned for use of the master stream appliance. In another scenario, the best plan of action might be

Figure 2-8 When the width and/or length of a building is involved, the ladder company must be positioned to accomplish the most important operations required of the aerial device and other equipment.

to position the ladder truck some distance from the building to allow fire fighters to perform their duties without unnecessary hazards.

When the fire building is one of several row buildings of the same height, the ladder truck can be positioned so that the roof of an adjoining building can be laddered, as shown in **Figure 2-9**. The fire fighters can then work their way from the adjoining roof to the fire building roof to perform ventilation, trench cut operations, or other ladder company tasks. This should not be done if the aerial device is required for rescue at the fire building.

Figure 2-9   When row structures are involved, a ladder truck can be positioned at adjoining buildings to get fire fighters to the roof.

Wrap-Up

Chief Concepts

- Ladder company tasks are performed at every fire. A ladder company is equipped with the tools and equipment needed to carry out these duties, and no member should leave the truck empty handed. Procedures should be devised to assign responsibility for tools and equipment carried on the truck.
- Company officers must form an initial sizeup are first to ensure proper positioning of the apparatus for efficient fireground operations. Ladder companies, as well as every other fire fighters at an incident, work within an incident management system. This system must have an overall strategic plan that outlines major objectives and provides the central focus for operations. The incident commander, whether the first-arriving company officer, or the chief of the department is responsible for maintaining the strategic plan as it changes throughout the incident.
- After arrival, ladder companies have several tasks to perform on the fire ground. Initial tasks include rescue, including the primary search and removal of occupants, forcible entry, and ventilation. Ladder companies must be well trained, adequately staffed, and supplied with modern fire apparatus and equipment enabling them to perform their assigned tasks in an efficient and safe manner.

Key Terms

Hand tools: Usually categorized as cutting, prying, or push/pull devices. These tools can be used to break locks, to force exterior and interior doors, to open sidewalk doors and grates, and generally to provide entry into a building.

Incident commander: Officer who is in charge of the incident and who develops a strategic plan for performing tasks during fireground operations.

Incident management system (IMS): The combination of facilities, equipment, personnel, procedures, and communications under a standard organizational structure to manage assigned resources effectively to accomplish stated objectives for an incident.

Power tools: Electricity, gasoline engines, or air or hydraulic pressure powers these tools.

1. The minimum staffing on a ladder company is
   a. 3
   b. 4
   c. 5
   d. An exact number of fire fighters is not given in the text, but the ladder company must be configured so that assigned personnel can perform all required tasks in as short a time as possible.

2. All crew members should be
   a. Assigned specific tools before arrival, but exact duties will vary according to conditions
   b. Assigned specific duties before arrival, but the tools will vary according to needs
   c. Assigned specific tools and duties before arrival
   d. Preassignment of tools and duties is seldom a good idea because of the many situations and conditions that could confront arriving companies.

3. Power tools ease and speed up the work of the ladder company; therefore,
   a. They are preferred when ventilating or forcing entry.
   b. They are preferred for forcing entry, but not always the best choice for ventilation.
   c. They are preferred for ventilation jobs, but are seldom a good choice for forcible entry.
   d. They are preferred for many tasks, but may be difficult to use in small areas.

4. Training opportunities are rare for
   a. Raising ground ladders
   b. Operating the aerial ladder
   c. Using ladder company tools
   d. All of the above

5. In situations where two ladder companies respond,
   a. The first-arriving ladder company should respond to the front of the building.
   b. The second-arriving ladder company should respond to the rear of the building.
   c. The second-arriving ladder company should position away from the building.
   d. The positions for the first- and second-arriving ladder company should be modified as needed.

6. When should a ladder company use the aerial device?
   a. It should be used as it is needed.
   b. It should be raised at every working fire.
   c. It should be raised at fires within reach of the aerial.
   d. All of the above

7. The ladder apparatus need not be positioned to use the aerial at
   a. One-story buildings
   b. Two- or three-story buildings
   c. Commercial occupancies
   d. All of the above

Rescue

Chapter 3

Learning Objectives

- Examine the chronology of a rescue operation.

- Identify several factors used to size up an incident to ascertain which rescue operations may be needed.

- Assess rescue considerations using several scenarios covering buildings of varying occupancies.

- Understand that a thorough planned search for victims should be conducted at every fire based on a risk-versus-benefit analysis.

- Consider different search patterns and techniques used during a primary search.

The primary search and the rescue of victims is the first and highest priority of fire fighters at an incident. It is, at times, the most challenging task to perform and may require great personal risk to those carrying out the task. To ensure the safety of all fire fighters committed to this objective, the following should be considered. A risk-versus-benefit analysis should be conducted to determine whether a search and rescue operation should be conducted. Command must be able to make an informed decision about whether it is safe to begin a search.

All efforts toward the search and rescue objective must be coordinated through command following department standard operating guidelines, approved procedures, and the overall strategic plan, which outlines the major objectives and provides the central focus for operations. There should be no freelancing allowed on the fire ground.

Fire fighters must be thoroughly trained and adequately staffed to perform search and rescue activities. In the initial stages of an incident, personnel must follow the two-in, two-out rule that requires a minimum of two personnel to enter a hazardous area and a minimum of two backup personnel to remain outside ready to rescue the fire fighters inside the hazardous area. A rapid intervention crew should be established as the incident progresses. The objective of the rapid intervention crew is to stand by with adequate personnel and equipment needed to provide emergency assistance to fire fighters working inside the hazardous area.

Fire fighters must be in full protective clothing including SCBA. They must have the proper tools and equipment to perform their task including portable radios, hand and/or power tools, lights, search rope, and thermal imaging devices.

Fire fighters must conduct a thorough, planned search using standard search patterns. They should be in teams consisting of two or more to safely carry out their objective. Multiple teams may be needed to conduct the search in a minimal amount of time depending on such factors as the type of occupancy, the size of the building, and fire conditions. Normally, occupants closest to the fire are in the most danger.

Command must ensure a coordinated fire attack between the engine and ladder company crews to make certain that tasks, such as fire suppression, are underway, that placement of hose lines for the protection of rescuers and victims are in place, and that ventilation is being conducted to support the search and rescue effort.

Locating and removing victims from a fire building are no easy tasks. Techniques for safely removing victims, including lifts, drags, and carries, should be taught and practiced. It is much easier and a lot safer to carry an unconscious child down a flight of stairs than to carry a large adult over a ground ladder.

Adequate personnel must be present to rescue conscious and/or unconscious victims as quickly as possible through the most advantageous escape routes available.

Search teams must notify the incident commander as to the progress and results of their efforts. The incident commander will adjust the overall plan based on the information received by those conducting the search for and rescue of victims. Decisions to continue or terminate the primary search then will be determined. In addition, the accountability of those conducting the search and rescue effort can be maintained. The safety of fire fighters working under these difficult conditions is paramount. When the risk outweighs the benefits of a successful search-and-rescue operation, for whatever reason, the danger to personnel may be too high to continue the operation.

Every fire fighter knows that the rescue of people in danger is the primary objective of a fire company and the first duty to be performed at the scene of a fire; however, not every fire fighter realizes that rescue duties involve much more than the bodily removal of people who might be trapped. Although carrying a fire victim to safety is rescue in the purest sense, rescue work includes many other operations.

Raising, climbing, and assisting entrapped occupants down a ladder, assisting or directing people from the fire building, and searching for victims in the building are all rescue operations, each of which immediately reduces danger to human life. Rapid ventilation removes accumulations of smoke and gases and prevents their further buildup. Proper placement of the first hose lines at a fire can keep the fire away from people in the building. Both operations reduce the danger to entrapped occupants and extend their time to get out of the building. In a very real sense, these too are rescue operations.

Many people tend to think of rescue only in connection with hospitals, nursing homes, schools, hotels, and other occupancies containing a lot of people. Such buildings must receive careful consideration in terms of rescue problems because of the number of people involved, but fires in one- and two-family dwellings very often require rescue operations. A review of national statistics shows that injuries and deaths in dwelling fires far outnumber those in other occupancies.

For ladder companies, rescue is a complex operation. Almost every rescue situation calls for a different combination of movements, equipment, and other fireground tactics. These can include forcible entry, interior search, placement of ground ladders, and ventilation.

This chapter discusses rescue work, beginning with preincident planning; search is considered in greater detail. The use of ground ladders, aerial devices, and forcible entry and ventilation procedures for rescue are discussed in later chapters.

Chronology of Rescue Operations

Before the Alarm

Preparation for rescue begins well before the alarm is received. It begins with building and area inspections and a continuing examination of the company's district to determine the occupancies, the people involved, the hazards, the potential rescue problems, and the most effective apparatus positioning. The objective is to know in advance the approximate type and extent of rescue operations that might be required at any fire.

At the Alarm

The information received with the alarm can be the first indication that a rescue problem exists. The initial information might include an exact address or a more general location such as a street intersection. From this information, ladder company personnel should know the type of occupancies involved or the type of area to which they are responding. This knowledge along with the day of the week and time of day are important clues to

> ### Key Points
>
> Preparation for rescue begins well before the alarm is received.

> ### Key Points
>
> The information received with the alarm can be the first indication that a rescue problem exists.

the possible presence of victims in the fire building and to the type of rescue problems likely to be involved **Figure 3-1**.

The information given with a verbal dispatch can be important. Such phrases as "across from," "next door to," "at the rear of," and "near the intersection of" can indicate that the alarm was not turned in by an occupant of the building, but by someone outside who saw smoke or flame coming from the building. People inside the building might be unaware of the fire or be unable to escape it at the time of the alarm.

> ### Key Points
>
> First reports, whether received by a central dispatcher or by someone receiving the information while on watch, should be relayed in their entirety to the company officer.

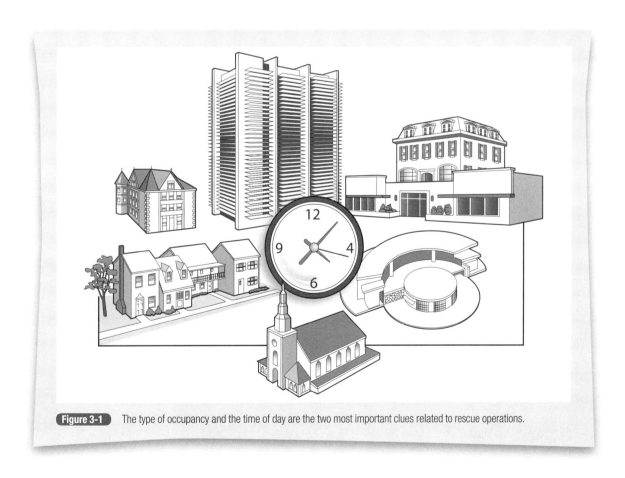

Figure 3-1 The type of occupancy and the time of day are the two most important clues related to rescue operations.

When dispatched by the emergency communications center, fire fighters should listen to and/or read all of the information pertinent to the incident. This will allow the responding companies to extract as much information as possible from the reports, including indications of any rescue problems.

On the Fire Ground

Sizeup   Even before the apparatus is stopped, the officer should have begun a careful sizeup of the situation. The officer should make visual observations to gather information to answer the following questions, which may vary depending on several factors, including the occupancy type.

- What type of occupancy is on fire?
- Is the fire building a closed-up dwelling with heavy smoke showing?
- Are cars parked in the driveway, front, or rear, indicating that an entire family might be inside?
- Are people at the windows of an apartment house, office building, or similar occupancy calling for help?
- In multiple occupancies, with smoke and/or fire showing, can calls for help be heard coming from the inside?
- If smoke and/or fire are showing, from where in the building does it originate?
- Given the interior construction of the building, in which directions will the fire travel most rapidly?

Reconnaissance and communication with other fire fighters, occupants, and neighbors or bystanders will assist in making critical decisions during the initial phase of operations. In addition, the extent of the fire, the size and age of the building, and its apparent number of occupants are important in ascertaining what rescue operations are needed. Some of this information would have been obtained in preincident planning and building inspections.

Of special urgency are reports that people are still inside. On the other hand, reports that "everyone is out" might be erroneous and should not deter fire fighters from beginning the primary search, especially in multiple-family residences. During sizeup operations, ladder crews must look for signs that people are in the building, as shown in **Figure 3-2**. Some signs that people are in the building include cars parked in front of the house or in the driveway, toys in the yard, or items on a porch, such as a bicycle. The sizeup will indicate where search-and-rescue operations should begin.

Figure 3-2   During sizeup, ladder crews must look for signs that people are in the building.

Immediate Rescue   **Immediate rescue,** at the expense of other ladder company operations, must be attempted in extreme cases, such as when an arriving company finds occupants about to leave the building from upper floors by jumping because of deteriorating conditions. In such situations, ladder company crews must delay all other operations in favor of raising ladders. They must get the attention of victims to make it obvious that rescue is underway and must talk to the victims to calm them until they can be brought down. Battery-powered megaphones are useful for this purpose.

The presence of victims at windows, needing help to get out, often indicates that other occupants are unable even to reach windows and are trapped within the building. When any single rescue operation keeps fire fighters from other duties, additional resources should be called for immediately.

Often, the sight of fire fighters arriving on the scene tends to calm panicky people. Otherwise, fire fighters must take immediate action to control overexcited occupants. One way to do this is to give positive orders and directions, in a forceful manner that will assure victims to follow your command. Hopefully victims will listen to reason and rescue can be made. This approach should also be followed when encountering victims during search operations.

Placement of Hose Lines   The stretching and water supply for hose lines are the tasks of responding engine companies. During rescue operations, hose lines are placed between the fire and the victims. This separates the fire from the people closest to it. Water streams from hose lines can be used to

- Control interior stairways and corridors for evacuating occupants and advancing fire fighters
- Protect fire fighters searching for victims around and above the fire

Hose lines should be placed as soon as possible on arrival to coordinate with the search efforts.

Search   If there is any indication that victims might be trapped or overcome within the fire building, a search should begin immediately. Search is discussed in detail in the last section of this chapter. Perhaps no other operation demands as much cooperation and coordination between ladder and engine companies. If ladder crews are to search around and above the fire, they must know that the fire is under attack and that attempts are being made to block fire spread. To use their hose lines effectively, engine companies might require ladder company operations, including ventilation, laddering, and forcible entry. These operations may also be required before the search can begin.

It is extremely important that every fire fighter at the scene be aware that a search is in progress. All activity should be directed toward assisting the ladder crews engaged in the search and providing protection for them and for any victims they may find. This coordinated fire attack is under the direct supervision of the incident commander.

Ventilation   The building should be ventilated as soon as possible to allow smoke and gases to move away from any occupants who could be trapped. This could be accomplished with horizontal ventilation directed from without or outside the building and/or vertical ventilation, performed from the top of the building.

Rescue Considerations

Rescue means searching for and removing victims and potential victims from danger. The extent of the rescue problem is directly affected by the following factors:

- Number of people in the fire building
- Paths by which fire and smoke can reach them
- Routes available to fire fighters for reaching people and removing them from the building

These factors, in turn, depend on the construction, size, and interior layout of the building. This section deals with rescue problems in several types of occupancies. It is important that the ladder company be properly equipped, adequately staffed, and well trained for efficient and safe rescue operations in the most complex structure in its district. As an extreme example, if an area contains only single-family dwellings, with the exception of one nursing home, ladder companies in that area must be as well prepared for rescue operations at the nursing home as they are for operations at the dwellings.

At any fire, but more often in larger occupancies, the rescue problem can be great enough to tax the capacity of the first

Key Points

It is important that the ladder company be properly equipped, adequately staffed, and well trained for efficient and safe rescue operations in the most complex structure in its district.

Key Points

In a typical two-story dwelling with one or two rooms on fire on the first floor, the occupants in most danger are those close to the fire on the first floor or those directly over the fire on the second floor.

units at the scene. If it is apparent or even suspected that such a rescue problem exists, additional resources should be requested without delay.

Following are several scenarios covering occupancies from single-family dwellings to a shopping mall. Fireground operations may vary among fire departments depending on their size, staffing levels, and equipment. What do not change among fire departments are the basic objectives conducted by ladder companies.

Residential Occupancies

Fires in residences of any type might require rescue operations at any time of day. At night, the rescue situation is usually more acute because occupants are usually home and could be sleeping. In addition, there are usually more people in a residence at night than during the day.

Single-Family Dwellings   In a typical two-story dwelling with one or two rooms on fire on the first floor, as shown in **Figure 3-3**, the occupants in most danger are those close to the fire on the first floor or those directly over the fire on the second floor. The former will be most affected by the radiant heat produced and the latter most affected by the convected smoke, hot air, and gases.

The main body of fire should be attacked immediately. As the attack on the fire progresses on the first floor, the area around the fire should be searched thoroughly. At the same time, ladder crews should be sent to the area over the fire to begin ventilating and searching for possible victims there. No time should be wasted in getting fire fighters to the upper floor. If an attack line is immediately available, it should be taken upstairs. If not, search should begin without it.

If fire is discovered in any upstairs room, that room should be searched if at all possible. Then the door should be shut to isolate the room until a hose line can be advanced. The doors and windows of other rooms should be opened to provide ventilation, dissipate heat and smoke, and allow more efficient search and fire attack operations.

In addition to the front entrance and stairs, access to the second floor can be by rear stairs,

Figure 3-3   A one- or two-room fire on the first floor of a single-family detached dwelling poses the greatest threat to the occupants close to the fire on the first floor or to those directly over the fire on the second floor.

Key Points

In any large, occupied building, the location of the fire and of most of the smoke above the fire should be carefully noted during sizeup. Smoke indicates the area into which the fire will most likely spread, the path it will take, and the location of occupants who will be most endangered if the fire does spread.

porches, and ladders. Search of the bedrooms, as well as all other rooms, and other ladder company duties such as forcible entry and ventilation can be carried out while engine company personnel advance hose lines on the floor below to control the stairway.

Apartment Houses   In any large, occupied building, the location of the fire and of most of the smoke above the fire should be carefully noted during sizeup. Smoke indicates the area into which the fire will most likely spread, the path it will take, and the location of occupants who will be most endangered if the fire does spread. In **Figure 3-4** , the victim at the upper right window may be in greater danger than the victim at the lower left window, even though the victim at the right is further from the fire. More victims are overcome by carbon monoxide than are burned. Therefore, while the fire is being attacked, every effort should be made to ventilate the building.

A search of the fire floor, the floor above the fire, the top floor, and then other floors in ascending order should be started as soon as possible to make sure that all occupants are located and removed from exposure to toxic gases. Here, even more than in a dwelling, search and rescue must be coordinated with a properly mounted attack on the fire.

Hotels and Motels   These residential occupancies present problems similar in some respects to those of large apartment houses. The rescue problems depend on the size, age, general construction, and population of the building; however, although there are usually fewer people per unit in a hotel or motel than in an apartment building or rooming house, there will be more units in the transient-type occupancy. Thus, as many people can be found in a hotel or motel as in an apartment building of the same size.

The number of people in a transient occupancy varies with the time of day and day of the week. Hotels, and especially motels, contain many more people at night.

Key Points

In the typical two- to four-story motel, the rooms directly behind those on fire must receive prime attention because of the possibility of lateral travel of smoke, gases, and fire between units.

When a working fire is encountered in this type of building, it must be thoroughly ventilated. The fire must be attacked, and the fire floor, the floor above the fire, the top floor, and then the other floors must be searched. In the typical two- to four-story motel, the rooms directly behind those on fire must receive prime attention because of the possibility of lateral travel of smoke, gases, and fire between units.

Industrial Occupancies   Factories, warehouses, and other industrial buildings present the greatest rescue problems during typical daytime working hours. Because offices are usually empty at night, daytime fires present the major rescue problem in these occupancies.

In many instances, however, a second shift works well into the night, and a third shift is employed for a 24-hour operation. To be properly prepared for all rescue situations, ladder companies assigned to industrial areas should be aware of the working hours in all such occupancies. Again, preincident planning and inspections will help in identifying any potential problems.

Another rescue consideration is the general physical ability of the employees. Many industrial buildings could employ handicapped or special needs workers who may be unable to exit the building without assistance.

Rescue can be hampered or complicated by burning chemicals or other hazardous materials. Large areas such as warehouses often require special rescue procedures to maintain control and contact with fire fighters engaged in search. Use of search rope procedures following a standard operating guideline should be in place for such an incident. Such variables affect the success of a potential rescue operation and point to the need for preincident planning and inspection.

Hospitals, Schools, and Institutions

Fires in hospitals, nursing homes, similar institutions, and schools are handled in essentially the same way as fires in multiple-family housing. Here, however, the search and rescue

Key Points

To be properly prepared for all rescue situations, ladder companies assigned to industrial areas should be aware of the working hours in all such occupancies.

Key Points

Rescue can be hampered or complicated by burning chemicals or other hazardous materials. Large areas such as warehouses often require special rescue procedures to maintain control and contact with fire fighters engaged in search.

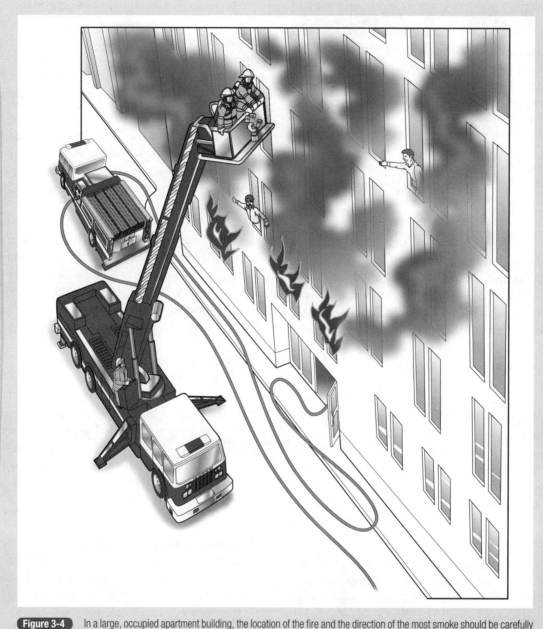

Figure 3-4 In a large, occupied apartment building, the location of the fire and the direction of the most smoke should be carefully noted, as those areas pose the greatest danger for occupants.

Key Points

Fires in hospitals, nursing homes, similar institutions, and schools are handled in essentially the same way as fires in multiple-family housing. Here, however, the search-and-rescue operation is compounded by a different group of people who may be young or old, nonambulatory, and physically and/or mentally challenged.

operation is compounded by a different group of people who may be young or old, nonambulatory, and physically and/or mentally challenged.

Although at night a hospital usually will have a smaller staff, no visitors, and a reduced maintenance crew, the smaller number of people does not make night rescue easier. Sleeping patients, sometimes sedated, often immobile, will be without the benefit of a full staff to assist them to safety. If the fire occurs in a vacant work area such as a kitchen, storeroom, or workshop, a small night staff might not promptly detect it; this could delay the alarm and increase the severity of the rescue problem. The same situation can be compounded in nursing homes by the high ratio of older occupants.

Special areas are often reserved for bedfast, nonambulatory patients and for patients who require continuing care. Fire companies must know the locations of such areas within hospitals and nursing homes in their district because these patients require extraordinary rescue procedures. Whether it is necessary to evacuate such an occupancy depends on its construction and size as well as the location and severity of the fire. If it is evident that the fire will be controlled, it could be possible to move patients within the building to areas that are isolated from the fire. This is true especially where corridors are divided by smoke barriers or fire barriers with appropriate smoke and fire control doors. Patient room doors are usually wide enough to accommodate a hospital bed or a mattress used as a stretcher to permit such movement.

Key Points

Special areas are often reserved for bedfast, nonambulatory patients and for patients who require continuing care. Fire companies must know the locations of such areas within hospitals and nursing homes in their district territory because these patients require extraordinary rescue procedures.

When smoke is the main problem, it might only be necessary to lower patients to the floor to get them below the smoke level, as shown in **Figure 3-5.** This can be done by lifting the mattress, with the patient on it, off the bed and placing it on the floor.

Either of these actions might be a desirable short-term solution in view of the physical condition of the patients, the continuing care some must receive, and possible adverse weather conditions; however, if there is any doubt regarding control of the fire, the building must be evacuated completely.

Most schools present a rescue problem in the daytime on weekdays from September to June; however, schools are often used after school, at night, and on weekends throughout the year for adult education, athletics, extracurricular activities, and other events. Every ladder company's responsibility is to be aware of such uses of the schools within its first-alarm response area.

After arriving at a working fire in a school, fire fighters should expect to find overly excited or panicky children and adults. The school should be completely evacuated. Areas around

Key Points

When smoke is the main problem, it might only be necessary to lower patients to the floor to get them below the smoke level.

Figure 3-5 According to the conditions, patients can be lowered to the floor or moved to an area clear of smoke.

and above the fire should be searched immediately if it is not certain that everyone has left the building.

Retail Stores

Consumers, for the most part, occupy stores during shopping hours. After hours, employees restocking merchandise may occupy stores. During certain selling seasons, shopping hours are extended, and stores are often overcrowded. Immediately before Christmas, for example, most stores are crowded with both people and stock. This is also true, to a lesser extent, before Easter, during school holidays, and before school starts in September. In all seasons, retail stores are most crowded on weekends and in the evenings.

Thus, the time of day, the day of the week, and the season all affect mercantile rescue operations. In addition, it should be noted that the occupants—both customers and employees—of most shops are women. Compounding normal rescue problems is the fact that many young children accompany the customers. The children sometimes leave their parents' side and roam through the store. A report of fire can quickly lead to mild panic as some adults search for their children and others rush for the exits.

Shopping malls, some covering acres of retail space and virtually hundreds of stores, contain many occupancy types, including restaurants, large chain stores, and places of assembly, including movie theaters. These malls can contain thousands of shoppers, and the life hazard can be high. A fire in a retail store could spread smoke throughout the area, causing panic

throughout the mall. Fortunately, most building regulations provide for buildings of this type to be sprinklered. A sprinkler system in any type of occupancy must be well maintained and operational to do the job it was designed to do: extinguish an incipient fire. During an actual fire in a mall, the incident commander may need to request additional resources to assist in the orderly and safe evacuation of the building.

Exits   In general, the larger a store, the greater the distance is between most customers and the exits. This makes it comparatively difficult for customers in large stores to get to an exit. Modern one-story "super stores," such as supermarkets, large drug stores, and discount stores, are laid out for restricted traffic flow. Narrow aisles between merchandise displays lead to even narrower checkout aisles and finally to only one or two exit doors. Turnstiles designed only to let people into the store, but not out, might block entrance doors. Shopping carts often are stored where they increase the difficulty of evacuating the building.

Door openings are usually limited in size in relationship to the floor area of the store and the number of people the store can hold. With fire and smoke closing in on customers, arriving fire fighters could find a panicky group of people pushing and shoving but unable to get out of the building. The primary goal of the fire department here is to get the people out. Ladder companies must evacuate the occupants and assist engine companies in getting into the store with attack lines.

Additional Openings   Ladder crews must calm the occupants and establish a traffic pattern out of the store. They must ensure that occupants move only in one direction—away from the fire—and that they will not interfere with others leaving or hamper the operation of fire fighters operating on the fire. People pushing on both sides can jam revolving doors; ladder truck crews must trip, or collapse, the doors to allow occupants to get out. If necessary, additional openings must be forced to allow occupants to leave and engine crews to enter.

In most modern retail stores, restaurants, and strip malls, storefront windows usually open directly to the sales floor. There is usually only 1 to 2 feet of exterior wall below the window, and thus, with the glass removed, these window openings can be used as additional exits for customers and entrances for engine crews. If possible, two separate windows should be removed. Customers should be directed to one opening, and engine crews should use the other to avoid congestion.

A plate-glass window must be removed carefully so that people inside or outside the store are not showered with broken glass. If possible, fire fighters should work from inside the store to knock out the glass toward the outside. The glass is first broken and knocked out at the top of the window with a pike pole, and then the rest of the window is cleared of glass from the top down. All glass must be removed from the window opening so that occupants exiting will not suffer a laceration from remaining shards of glass. Fire fighters should now enter the store and direct occupants to the opened window, away from the opening made for engine company personnel, as shown in **Figure 3-6**. Glass should also be removed from in front of the window. Glass lying on the ground presents a slippery surface in which a fleeing occupant may fall and receive a serious injury.

Search   After the evacuation or simultaneously if adequate resources are available, all areas of the building that are tenable must be searched. Retail space, offices, storerooms, and other areas must be searched. Occupants and employees may have sought shelter in these areas or might have collapsed or fallen on the way out. As with multistoried residences, the upper floors of multistoried stores must also be searched. The search is generally simplified by the lack of interior partitions.

Truck crews who through preincident planning have come to know the construction and layout of the stores in their district can carry out evacuation and search procedures most efficiently. As noted previously, the locations of windows and the operating mechanisms of entrance and exit doors are important in rescue situations.

Search

A thorough, planned search for victims should be conducted at every fire. Moreover, all fire fighters, no matter what type of company to which they are assigned, should be conducting a search as they move through the building. All personnel should realize that the safety of fire fighters engaged in search is their responsibility.

Key Points

If possible, two separate windows should be removed. Customers should be directed to one opening, and engine crews should use the other to avoid congestion.

Key Points

After the evacuation or simultaneously if adequate resources are available, all areas of the building that are tenable must be searched.

Figure 3-6   Store windows can be used for moving out occupants and moving in fire fighters.

Key Points

A thorough, planned search for victims should be conducted at every fire. Moreover, all fire fighters, no matter what type of company to which they are assigned, should be conducting a search as they move through the building.

Crews operating a hose line on the fire floor must be aware that ladder crews will be searching the floors above for possible victims. If those on the hose line cannot control the fire and are forced to retreat from the building, ladder crews above must be warned so they can take appropriate action. Fire fighters assigned to laddering duties also must be aware that a search is in progress, as they might have to place ladders as additional exits from upper stories for fire fighters performing the primary search and possible victims.

Because search is an overall responsibility, it should be performed according to an efficient standard operating guideline and coordinated by command. Fire-fighter safety is paramount during the search, and both the engine and ladder companies must coordinate their efforts for a successful operation. The procedure should be simple and straightforward so that one fire fighter can substitute for another at any point in the search. It should include an easy-to-follow search pattern that all fire fighters are familiar with. The primary search should begin where occupants are in the greatest danger.

Normally, occupants closest to the fire are in the most danger, whether they are on the fire floor or the floor above. Radiant heat, smoke, and fire gases will affect those on the fire floor, whereas those above will be subjected to convected heat, fire gases, smoke, and hot air. Occupants who are two or more floors above the fire (and especially on the top floor) can be endangered by smoke and heat that have been channeled up through the building; therefore, ladder crews must begin both search and ventilation operations quickly, while engine crews attack the fire.

Search Duties

It is important that a number of operations be carried out simultaneously, or as nearly so as possible, in any rescue situation. While following a standard search procedure, truck crews should perform the following duties, the first being by far the most important:

- Locate and remove trapped occupants.
- Ventilate where needed as they move along.

- Temporarily prevent extension of fire by closing doors and windows.
- Check for interior and exterior fire extension.
- Help locate the seat of the fire when necessary.

Standard Search Procedure

The best way to see the value of a standard search procedure and search pattern is to look at a typical search. This section creates a scenario with a fire underway in a large kitchen and dining area on the first floor of a two-story dwelling. Standard operating guidelines and the search pattern described apply to all buildings.

Immediate Search   Companies arriving at the scene should immediately size up the fire situation. Engine companies obtain a water supply and advance attack lines into the house. The attack lines are deployed to cut off the forward progress of the fire and then to hit it directly. Fire fighters on the attack lines, by getting low, can see some clear area above the floor, and they check for victims near the fire. The stairway and the upper floor are full of smoke and gases. An immediate attempt should be made to get ladder crews to the upper floor. If the area is tenable, they can begin searching for victims, as shown in **Figure 3-7** . If the area is untenable because of the intense heat, ventilation must begin from the outside.

Windows on the lee side of the dwelling should be taken out first. One fire fighter with an appropriate ground ladder can quickly knock out enough second-floor windows to make a big difference.

Search Pattern   As soon as the second floor is tenable, the primary search begins. According to standard operating guidelines, the most dangerous area—directly over the fire— is the first place to be searched. After reaching the top of the stairway, fire fighters will turn in one direction or the other to get to the room over the fire. They will begin the search by turning either right or left into that room. This turn sets up the basic search pattern for the entire floor. As the search proceeds, fire fighters performing the primary search will keep turning in the original direction as they go in and out of rooms.

Thus, if the first turn into the room over the fire is a right-hand turn, then all turns into subsequent rooms on the floor will be right-hand turns, as shown in **Figure 3-8a** . If the first turn into the room over the fire is a left-hand turn, then all turns into subsequent rooms will be left-hand turns, as shown in **Figure 3-8b** .

Key Points

Because search is an overall responsibility, it should be performed according to an efficient standard operating guideline and coordinated by command. Fire fighter safety is paramount during the search, and both the engine and ladder companies must coordinate their efforts for a successful operation.

Key Points

An immediate attempt should be made to get ladder crews to the upper floor. If the area is tenable, they can begin searching for victims. If the area is untenable because of the intense heat, ventilation must begin from the outside.

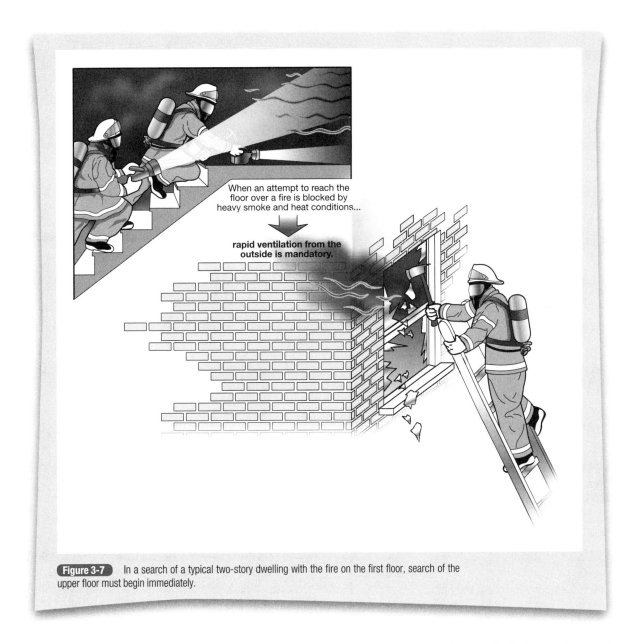

When an attempt to reach the floor over a fire is blocked by heavy smoke and heat conditions...

rapid ventilation from the outside is mandatory.

Figure 3-7 In a search of a typical two-story dwelling with the fire on the first floor, search of the upper floor must begin immediately.

For example, assume that fire fighters engaged in the search turn right at the top of the stairs and turn right again to move down the hallway to get to the room over the fire. Kicking off the search pattern for the floor, the fire fighters turn right to get into the room over the fire, as shown in Figure 3-8a. When they have finished searching that room, they turn right out of the room and begin the search of the hallway until they reach the next room to be searched. Then, according to the search pattern, the fire fighters turn right again to enter a room and, after coming out of that room, again turn right and move along the hallway to the next room. The right-turning path is continued until the fire fighters have worked all of the way around the hall and are back at the stairs.

Others searching the area should follow the same pattern. Whenever possible, at least two fire fighters should be assigned to search in an area. Depending on the circumstances, one fire fighter takes the first room found, and the other moves on to the next room. Another variation would have one fire fighter search the room while the other monitors conditions in the hallway and acts as a beacon to the fire fighter searching the room. Yet another variation has both fire fighters searching the room. This may be necessary if the room is large and the center of the room needs to be searched. A fire fighter could stay against the wall while the other moves away to the center, conducts the search, and then moves back to the wall.

Whichever search method is used, it must be accomplished thoroughly but with an urgency in mind. Under heavy smoke and/or fire conditions, victims must be found as quickly as possible to ensure the possibility of survival. As fire fighters

Key Points

Whenever possible, at least two fire fighters should be assigned to search in an area.

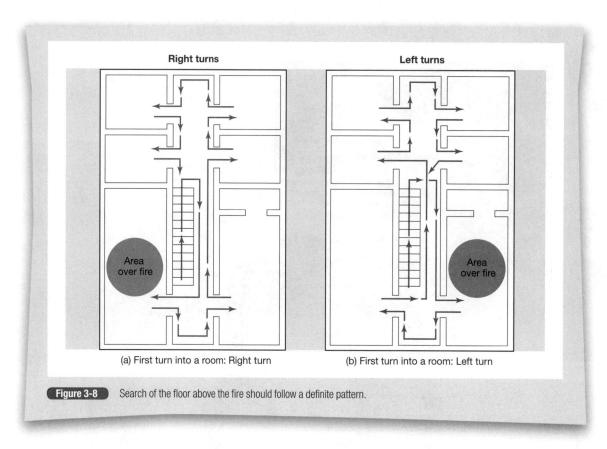

Right turns

Left turns

Area over fire

Area over fire

(a) First turn into a room: Right turn

(b) First turn into a room: Left turn

Figure 3-8   Search of the floor above the fire should follow a definite pattern.

proceed with the search pattern, they should check each room that is not marked as having been searched, as will be discussed later. They should keep track of each other by touch, sight, verbally, and by listening for the sound of breathing from the other's SCBA. Each should be alert for sounds from a victim, including a call for help from the victim or another fire fighter.

If fire fighters conducting the search need to reverse direction, they should reverse the travel of their search. The fire fighter who entered and went left-left-left would turn around and proceed right-right-right.

Areas to Be Searched   Corridors and halls should be checked thoroughly, as should the open area of each room. In addition, bathrooms, closets, and the spaces behind large chairs and behind and under beds should be checked, as shown in **Figure 3-9**. People, especially children, often seek protection in these places. Areas near windows should be checked for victims overcome while attempting to reach a window. Everyone engaged in the search should carry a tool such as an axe, a Halligan bar, or a claw tool, with which these areas can be probed and which can be used for venting and forcing locked doors. Each person

making a search should also carry a hand light as well as a bag of search rope. In addition, all SCBA should be equipped with an integrated **personal alert safety system.**

If the fire has extended into a room, that room should also be searched if at all possible. The door to the fire room should be shut to isolate the fire **Figure 3-10**. Standard procedures for engine companies should include a hose line over the fire for just this situation. With a charged hose line available, there is a better chance for the search to proceed quickly and safely.

If a room is not involved with fire but contains heat and smoke, it should be vented. Its door should be opened, and its windows opened or glass removed to clear the area for a more effective search. Windows in hallways and corridors should be opened or removed.

Indicating that a Room Has Been Searched   Establish a standard way of indicating that a room has been searched so that there is no duplication of effort, at least not in the initial search operation. When the door to a room is to be left open to vent through the room, an effective indication is to place a piece of light furniture in the doorway. A chair, footstool, end

Key Points

If the fire has extended into a room, that room should also be searched if at all possible. Then the door to the fire room should be shut to isolate the fire.

Key Points

If a room is not involved with fire but contains heat and smoke, it should be vented. Its door should be opened, and its windows opened or glass removed to clear the area for a more effective search.

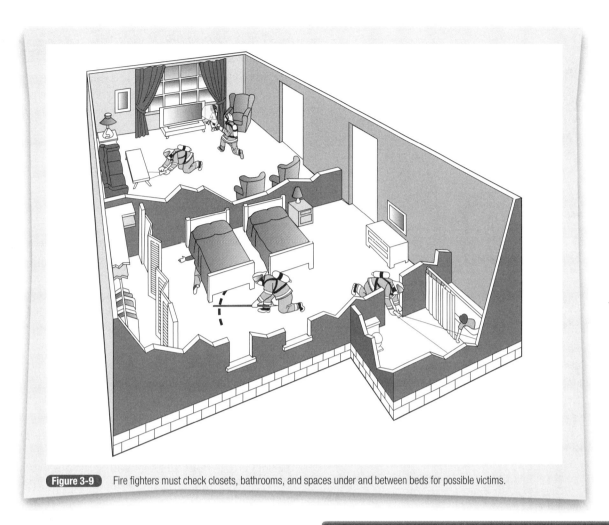

Figure 3-9 Fire fighters must check closets, bathrooms, and spaces under and between beds for possible victims.

Figure 3-10 When finding that fire has entered a room, the room should be searched, if possible, and the door shut to isolate the fire.

Key Points

Establish a standard way of indicating that a room has been searched so that there is no duplication of effort, at least not in the initial search operation. When the door to a room is to be left open to vent through the room, an effective indication is to place a piece of light furniture in the doorway.

table, lamp table, or anything else that can be quickly dragged or carried into position will do. This is quickly done and easily recognized by other search personnel.

If fire has extended into a room or if there is danger of this happening, the door must be closed after the room is searched. The door should be marked to indicate that the room has been searched. This can be done by drawing a large "X" on the door with chalk or hanging a tag on the door knob, as shown in **Figure 3-11**. A piece of cloth could also be attached to the doorknob or placed against the doorjamb, at about doorknob height, and the door pulled shut against it. The corner of a bed sheet, a towel, or pillow case, a tablecloth, or an item of clothing can be used for this purpose. In a situation where fire has entered the room, an attack line should be called for immediately.

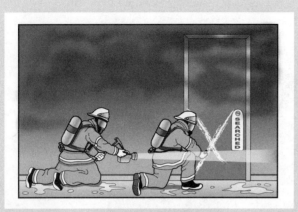

Figure 3-11 Hanging a tag on the door knob or marking the door with a large "X" will indicate that the room has been searched.

Key Points

If fire has extended into a room or if there is danger of this happening, the door must be closed after the room is searched. Then a piece of cloth could should be attached to the doorknob or placed against the doorjamb, at about doorknob height, and the door pulled shut against it.

Some departments have had success with placing commercially available tags on the doors of inspected rooms, apartments, and offices. Chalk and inner tube rubber are inexpensive alternatives for marking searched areas. Whichever type of search indication is used, fire fighters need not duplicate the search process by entering a room to find that it has already been searched.

Other Structures   The standard search procedures and the search pattern described here apply to other buildings as well. The search pattern may be used in a single-story dwelling, an apartment house, or an office building.

After they turn off a corridor into an apartment or office, the fire fighters engaged in the search should follow the same pattern within the room or area. When they leave the area, they should retain the pattern in the corridor and when they enter the next room or area. If they enter an office area with a right turn, they place their right side to the wall and must then keep their right side to the wall as they work their way through the area and back to the door leading to the hall. If they enter with a left

Key Points

After they turn off a corridor into an apartment or office, the fire fighters engaged in the search should follow the same pattern within the room or area.

Key Points

It is important that search personnel leave an office or work area through the same doorway used to enter it. Otherwise, a part of the area that could possibly contain trapped victims could be overlooked.

turn, they should keep their left side to the wall as they search the area and work their way back to the door. When they leave the area, they must retain this pattern in the hall and in entering the next office area, as shown in **Figure 3-12** .

It is important that search personnel leave an office or work area through the same doorway used to enter it. Otherwise, a part of the area that could possibly contain trapped victims could be overlooked **Figure 3-13** . If fire conditions force search teams to leave by a different door, they should report this information to command immediately. Hose lines should be quickly advanced to the area thoroughly searched.

Large structures or buildings with complex floor plans will require the use of additional resources. Large office areas with cubicles are labor intensive to search properly under heavy smoke conditions. Obstructions of any kind in unfamiliar surroundings under fire conditions add an additional burden to fire fighters conducting the primary search. Command must ensure adequate personnel to perform all fireground operations. Resources must be available to perform the primary search with deliberate purpose and in a safe manner. Members will have to be relieved and additional fire fighters assigned to continue the search.

Search and rescue are ladder company duties on the fire ground. Engine companies should be free to direct all their efforts to advancing hose lines in support of rescue operations; however, where ladder companies are overburdened, do not have adequate personnel, or do not respond in sufficient numbers, engine company personnel must then be assigned to perform search-and-rescue operations. With standard operating guidelines and proper training in place, assignments can be adjusted to cover any situation.

Search Techniques

Search personnel must work as quickly as possible, but they must also be careful to avoid unnecessary injury to themselves and victims. This is especially true when searching near or above the fire. Fire fighters must check for extension of fire into areas where they may be working. Fire may be in adjacent rooms hidden behind closed doors or in concealed spaces such as walls and ceilings.

Key Points

Large structures or buildings with complex floor plans will require the use of additional resources.

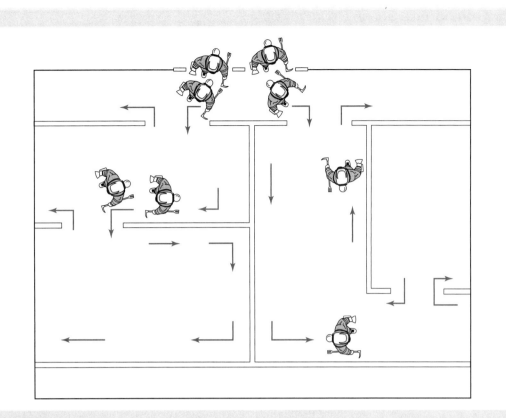

Figure 3-12 When turning into an office, apartment, or similar area, fire fighters should keep one shoulder to the wall: left if a left turn is made upon entering, right if initial turn is to the right.

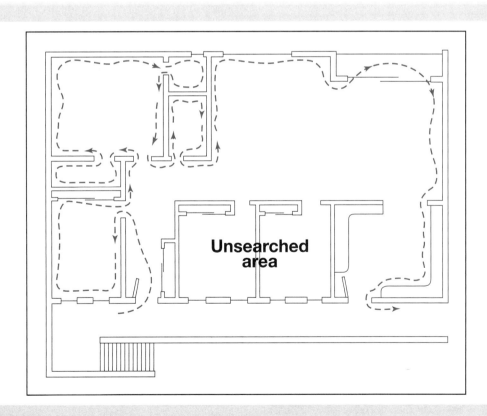

Unsearched area

Figure 3-13 During search operations, the same door should be used to enter and exit an area in order to prevent missing some of the rooms.

Doors   Before any door is opened, it should be checked to see if it or the doorknob is hot. A very hot knob usually indicates that there is fire on the other side of the door. Some heat in either the knob or the door probably means that the room beyond the door is full of smoke and gases, but not fire. The gases could ignite suddenly and explosively if the door is opened quickly. In such situations, the safest course is to first determine which way the door opens. Most doors between hallways and apartments or offices open into the unit; that is, away from the entering fire fighter. Inside the unit, doors may open in either direction. If the hinges are on the outside, then the door opens out toward the fire fighter. Some modern doors with flush, hidden hinges give no indication of which way they open; however, if the doorjamb does not cover the edge of the door, then the door opens out. If the edge of the door is covered, it opens into the unit.

Doors that open out are the most dangerous to fire fighters, but they must be opened during the search. If an outward-opening door shows some heat, the fire fighter should get low and place his or her full body weight against the door. Then he or she should release the lock slowly and allow the door to open slightly. As the lock is released, the fire fighter may feel a strong push against the door from the inside. This indicates that the room is full of smoke and gas, possibly along with fire. The fire fighter should close the door immediately and call for an attack line, and the door should not be opened again until the hose line is in position. The same action should be taken when there is no push against the door, but the room is seen to be full of fire.

If the door opens into the room, the lock should be released slowly and the door eased in. Again, if there is a strong push against the door or if the room is heavily involved with fire, the door should be shut and an attack line brought into position. Otherwise, the unit or room can be entered and searched. If doors are locked and need to be forced open, caution should be used following the information provided previously here.

Victims   Occupants of a fire building will try to escape through doors, windows, fire escapes, hallways, and stairways leading out of the building. Fire fighters should look for victims who may have been overcome in these areas. In particular, fire fighters

must be sure that when opening an inward opening door a victim is not lying behind it. If the victim is near the door, it may open easily at first, and the fire fighter will feel a bump as the door hits the victim.

If there is fire within the room, an attempt should be made to open the door enough to remove the victim and then shut the door. Just getting the victim out of a smoke-filled room and into a less-charged hallway might mean the difference between life and death. If necessary, additional help should immediately be called to assist in removing the victim from the area. If an attack line is needed, it should be advanced to the location, and the unit should then be searched for other victims.

When a door opens inward easily at first and then bumps an object, the fire fighter must assume that there is a victim just inside the door. Few people place their furniture or other possessions behind a door so that it blocks proper egress. The door must be opened, but the fire fighter should not force the door, because this could only add to injuries as the door is slammed into the victim. Instead, the fire fighter should drop down and feel around the inside or probe with tools to find the victim. If possible, get the victim away from the door. Open it, and remove the victim from the room **Figure 3-14** .

Great physical effort is often required to move a victim away from a door. The fire fighter must work from an awkward position with a door in the way and must move a dead weight. If help is needed, call for it quickly.

If enough fire fighters are available, victims should be removed from the building while the search continues; otherwise, victims should be removed and then the search resumed.

Figure 3-14 When a door will open only partway, a fire fighter should not attempt to force it open further in case a victim is blocking the door.

If the victim is found deep within an apartment or a large room or office, the best action is still to move the victim out of the building as quickly as possible. If it is impossible to do this, it may be advantageous to move the victim into an adjoining apartment or room while awaiting additional resources to move the victim. In extreme circumstances, fire fighters may have to breach a wall or partition to move the victim into an uninvolved area of refuge **Figure 3-15**.

Visibility When smoke reduces visibility during search operations, fire fighters should stay low and move quickly on hands and knees. The general rule should be as follows: "If you can't see your feet, you should be on the floor." The hands and lower legs should be used to feel for victims. Body weight should be kept to the rear and a tool should be used to sound the floor and check for holes in the floor, open shafts and obstructions. A personal rescue rope is a valuable asset in areas of reduced visibility. The rope can be attached to a permanent fixture such as a doorknob, radiator, or pipe. As the fire fighter advances, the rope plays out of a bag attached to the SCBA. Search rope may be of any length, but common lengths are 40 to 50 feet. The rope is attached to the rope bag, and thus, the fire fighter cannot advance further than the length

Figure 3-15 In extreme situations or when deep in a building, it maybe advantageous to force walls or partitions to get victims into a less-charged area or to a window.

of the rope. If the fire fighter becomes disorientated, reversing direction by following the rope will lead back to the starting point.

In hallways and corridors, the walls can serve as directional guides. As search personnel feel along hallway walls, the locations of doors will become obvious. Inside the rooms of apartments and office suites, the hands-and-knees position will keep fire fighters from tripping over furniture and other obstacles and will help in locating victims.

Windows should be opened or removed as they are encountered. This helps relieve the visibility problem and makes it easier for search personnel to find and reach victims. It also reduces the danger from heat, smoke, and gases to fire fighters and victims alike. Ventilation techniques and operations are discussed in detail in the next two chapters.

Key Points

If the victim is found deep within an apartment or a large room or office, the best action is still to move the victim out of the building as quickly as possible. If it is impossible to do this, it may be advantageous to move the victim into an adjoining apartment or room while awaiting additional resources to move the victim.

Key Points

When smoke reduces visibility during search operations, fire fighters should stay low and move quickly on hands and knees. If you cannot see your feet, you should be on the floor. The hands and lower legs should be used to feel for victims.

Wrap-Up

Chief Concepts

- Rescue is the prime objective of fire companies and the basic reason for their existence. Fire fighters must expect some personal risk during rescue operations. To minimize this danger, the search must be conducted following standard operating guidelines under the direction of the incident commander. Fire fighters should receive continual training in these operations, should possess the proper equipment, and should be adequately staffed to carry out this operation safely and effectively. These procedures should include a standard search pattern that is simple to perform as well as thorough in its coverage of the fire building.

- Search, ventilation, and forcible entry are the main functions of ladder companies in a rescue situation, along with laddering the building. Although rescue operations are usually assigned to ladder company personnel, it must be remembered that all fire fighters are committed to the search. When ladder companies are not available, other fire fighters on the fire ground will be required to conduct all aspects of a primary search, including forcible entry, ventilation, and providing egress from the building.

Key Terms

Immediate rescue: This action must be attempted in extreme cases, such as when an arriving company finds occupants about to leave the building from upper floors by jumping because of deteriorating conditions. In such situations, ladder company crews must delay all other operations in favor of raising ladders.

Personal alert safety system: Device worn by a fire fighter that sounds an alarm if the fire fighter is motionless for a period of time.

1. The greatest number of deaths and injuries due to fire occur in
   a. Hotels
   b. Hospitals
   c. Places of assembly
   d. Single-family dwellings

2. Which of the following are important factors in determining what rescue operations are needed?
   a. Extent of fire
   b. Size and age of the building
   c. Number of occupants
   d. All of the above

3. When occupants are seen at windows, this means that
   a. There may be other building occupants who cannot be seen and who are trapped.
   b. These occupants are in the most danger.
   c. All victims have either self-rescued or made it to a window.
   d. All other tasks should be delayed until all visible occupants are removed.

4. Ladder companies must be properly equipped, adequately staffed, and trained to handle fires in the types of buildings and occupancies in their response area that
   a. Are most abundant
   b. Contain the largest number of occupants
   c. Present the greatest challenge in terms of height and area
   d. Are the most complex

5. In residential occupancies
   a. Rescue operations may be needed at any time of day.
   b. Rescue operations tend to be a greater challenge at night.
   c. There tend to be more people at home at night.
   d. All of the above

6. Which of the following occupancies is most likely to require a search rope procedures?
   a. Single-family detached dwelling
   b. Individual units in an apartment building
   c. Patient care area in a hospital
   d. A production area in an industrial plant

7. When encountering smoky conditions in a hospital room, the text recommends which of the following as a short-term rescue method for immobile patients?
   a. Moving the entire bed through the door (doors leading to patient rooms are generally wide enough to move the entire bed through the opening)

b. Lower the patient to the floor by lifting the mattress with the patient on it onto the floor
   c. Lowering patients to the floor and removing them using a blanket drag
   d. Using one of the standard hospital carries shown in basic firefighting manuals

8. Searching for victims at a fire is the responsibility of
   a. The first-arriving ladder company
   b. The first-arriving engine company
   c. The second-arriving ladder company
   d. All companies as they move through the building

9. Of the duties that ladder company crews should perform, which of the following is by far the most important?
   a. Locate and remove trapped occupants
   b. Ventilate
   c. Temporarily prevent extension by closing fire doors and windows
   d. All of the above are equally important and should be carried out simultaneously.

10. When conducting a primary search in a room, the room should be immediately vented if
    a. There is a fire in that room.
    b. The room contains heat and smoke but no fire.
    c. The room contains fire or any products of combustion.
    d. Venting should always occur concurrently with the search.

11. When conducting a primary search in a room, the door to that room should be shut to isolate the fire if
    a. There is a fire in that room.
    b. The room contains heat and smoke but no fire.
    c. The room contains fire or any products of combustion.
    d. The fire should be isolated any time there is a fire separation available.

12. When search-and-rescue staffing is limited and priorities must be established regarding removal of victims from the building
    a. Immediately remove all victims to the outside.
    b. Immediately move all victims to either the outside or to an area separated from the fire area by fire-resistive construction with rated opening protective devices.
    c. Move victims from the room being searched to the hallway or if in a large room move the victim to an adjoining apartment or room, and resume the search.
    d. Move victims to a location near the door and call for assistance to remove them to the outside.

Ventilation Techniques

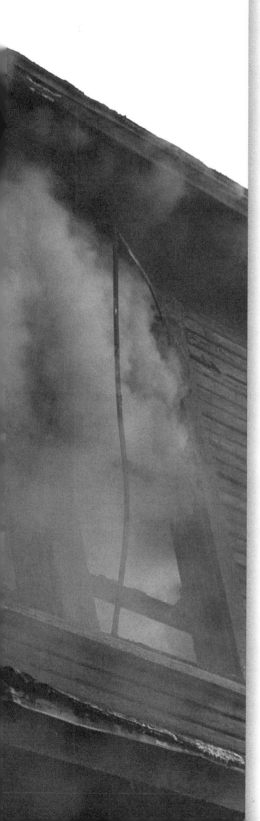

Chapter 4

Learning Objectives

- Define the term ventilation as it applies to the fire service.

- Identify ventilation factors that contribute directly to the accomplishment of basic firefighting objectives.

- Examine the basic principles of ventilation.

- Consider techniques for ventilation using a variety of natural openings in a building.

- Consider techniques for ventilation using tools to cut a hole in the roof.

- Consider techniques for ventilation using forced or mechanical equipment.

Ventilation, if performed properly, will allow heat and the products of combustion to leave the building, making conditions inside tenable for both occupants and fire fighters. Ventilation increases visibility and lowers temperatures that enable fire fighters to get inside to perform search-and-rescue operations as well as fire suppression activities in a safer working environment. They will be able to search the building quickly and more thoroughly. Fire fighters will be able to locate the seat of the fire, extinguish it, and check for extension. Proper ventilation will also reduce or eliminate the possibility of flashover or a backdraft explosion.

The safety of fire fighters and occupants could be put at risk by fire spread through the building if improper ventilation techniques are employed on the fire ground. As in all fireground operations, command must maintain a coordinated effort between ladder company and engine company personnel to conduct a successful fire attack.

Ladder trucks have been designed with the task of ventilation in mind. They contain an aerial device, ground ladders, and tools and equipment that allow fire fighters to reach areas of a building, from both inside and outside, where ventilation operations should take place.

Ventilation is the process of making openings in a fire building or exposure to allow heat and the products of combustion to leave the building.

Ventilation contributes directly to the accomplishment of basic firefighting objectives by

- Reducing the danger from heat and smoke to trapped occupants and extending the time available for fire fighters to perform search and rescue operations
- Increasing visibility, for both fire fighters and occupants, thereby decreasing the danger inherent in fireground operations and increasing fireground efficiency
- Permitting quicker and easier entry and allowing fire fighters to conduct search operations and advance attack lines into the building
- Minimizing the time required to locate the seat of the fire
- Minimizing the time required by ladder crews to locate and expose areas in which the fire has extended within the building
- Decreasing or stopping the extension of fire
- Reducing the chance of flashover or a backdraft explosion

Several factors will determine the results of ventilation in a fire building. These factors include the size and type of occupancy involved, the extent and location of the fire, and whether the fire is free burning or smoldering. The number of personnel assigned to the ventilation operation and how well they are trained can also impact its effectiveness; however, when properly performed, ventilation will increase the effectiveness of most fireground operations by increasing visibility, removing heat and gases and allowing fire fighters to enter the building **Figure 4-1**.

Key Points

The benefits of ventilation far outweigh its disadvantages.

In spite of the benefits of ventilation, many fire departments hesitate to ventilate buildings that are on fire. They note that ventilation techniques require doing damage to a building, whereas one goal of fire departments is to limit damage. The fact of the matter is that if done properly the small amount of damage done during ventilation operations ultimately results in a much larger reduction in fire damage. Even more important, ventilation directly aids in attaining the primary objective of firefighting operations: saving lives.

The benefits of ventilation far outweigh its disadvantages. The principle is simple and straightforward, and the techniques are easy to learn and use. This chapter deals with the methods by which fire buildings can be opened for ventilation. Chapter 5 discusses the use of these methods in ventilation operations.

Basic Principles

In Chapter 1, it was explained that fire travel by convection presents the greatest firefighting problem. It is by convection that hot air, smoke, heated gases, and burning embers move through a building, traveling vertically where possible and spreading horizontally when vertical pathways are blocked. If the horizontal pathways also are blocked, products of combustion build up and begin to fill the building from the top down. The products of combustion limit visibility within the fire building, cause death by asphyxiation, and ignite secondary fires that can be more severe than the original fire.

Figure 4-2 shows a multistory building with a working fire on a lower floor. The fire floor is filled with hot smoke and gases, but these combustion products have also been convected up from the fire floor through natural channels in the building. They have accumulated under the roof, which blocks their vertical pathways, and then have moved horizontally. The appearance of such a phenomenal occurrence has led to the use of the term "**mushrooming**" to describe this condition.

As the fire burns, hot air and combustion products will continue to rise to the roof. More and more heat will be pumped

Key Points

The products of combustion limit visibility within the fire building, cause death by asphyxiation, and ignite secondary fires that can be more severe than the original fire.

Figure 4-1 Ventilation, if carried out properly, will increase the effectiveness of fireground operations.

up, and eventually something will ignite. There will then be two fires in the building, possibly separated by several uninvolved stories.

To prevent the secondary fire from igniting, fire fighters must reduce the accumulation of the products of combustion under the roof by making an opening through which these products can escape. To make use of the natural tendency of heated products to rise, this opening should be above the accumulation of heat and gases. At least one ventilation opening of adequate size should be made at the top of the building directly above the fire where the situation is most acute, as shown in **Figure 4-3**. In other

situations, openings should be made at the tops of the vertical shafts through which products of combustion are rising.

Simultaneously, the accumulation of heat and the products of combustion on the fire floor should be relieved to prevent these toxic elements from moving through the rest of the building. Openings made on the fire floor are usually made through windows and doors that lead directly from the accumulations of these products to the outside of the building. Similar action should be taken on the floor above the fire, as soon as possible, so that the area can be searched and hose lines advanced to check for fire extension.

Key Points

To prevent the secondary fire from igniting, fire fighters must reduce the accumulation of the products of combustion under the roof by making an opening through which these products can escape.

Key Points

This is the general rule behind all ventilation operations: Open the fire building in such a way that all accumulations of heat and the products of combustion will leave the building by natural convection.

Figure 4-2   In a multistory building, products of combustion convected up from a fire on a lower floor create a condition known as "mushrooming."

This, then, is the general rule behind all ventilation operations: Open the fire building in such a way that all accumulations of heat and the products of combustion will leave the building by natural convection. The principle is obvious; the results, in terms of the reduction of deaths, injuries, property damage, and fire spread, are immeasurable.

Natural Openings

Opening the roof is the only way to reduce accumulated combustion products; to prevent them from collecting, there should be no hesitation in making the opening, but this is not always necessary. In many cases, a building can be effectively

Figure 4-3   A ventilation opening of adequate size should be made directly above the fire to help control fire spread and allow fire fighters to approach the fire and apply water to the burning areas.

ventilated through natural openings—"built-in" construction features that can be quickly opened and easily repaired. Windows are natural openings, as are skylights, roof hatches, ventilators, and penthouses. The effectiveness of using natural openings for ventilation depends on their location in relation to the fire and on the pathways open to the products of combustion.

Obviously, not all of the natural openings discussed in the following pages are found on every building. Ladder companies should through preincident planning and inspection determine which are available and how they can best be used for ventilation. Ladder crews should be able to recognize natural roof openings and the building areas served by these openings and should know the most efficient methods for uncovering them with standard hand and/or power tools.

Windows

Along with the location of the fire, the construction, type, and size of the fire building determine how it should be vented. What may be required to properly ventilate one building might be unnecessary in another. For example, unless the fire is in the attic, roof ventilation is rarely necessary in one- or two-family dwellings because these structures can usually be vented using horizontal ventilation through windows and doors.

When time permits, the windows should simply be opened. Double-hung windows should be opened about two thirds down from the top and one third up from the bottom. Other types of windows should be opened as much as possible. If a window is equipped with a storm window, it, too, must be opened conventionally or broken.

Shades, venetian blinds, drapes, curtains, and other window coverings must be moved away from the window. If they cannot be raised or moved to the side quickly, they should be pulled down and removed. If there is not enough time to open windows and storm windows, they should be broken with a tool suitable to perform the task. Use as little force as possible so that no broken glass will endanger fire fighters who may be working outside. Fire fighters outside the fire structure must be wary of the possibility of flying glass. Fire fighters performing ventilation from within a building should always notify fire fighters working outside of their intention.

In some cases, ladder crews cannot get inside the building to open windows. If this is the case, the windows should be broken from the outside with appropriate tools. Ground ladders or aerial devices may need to be raised to allow fire fighters to perform these tasks. These operations are covered in Chapters 8 and 9.

Effects of Wind   When wind is a factor, the windows on the leeward side of the building should be opened first, as shown in **Figure 4-4**. Then the windows on the windward side can be

opened to allow the wind to blow the products of combustion out of the building. If the windward side is opened first, the wind will churn smoke and gas around the interior until the leeward side is opened, but opening the windows in proper sequence will create effective cross ventilation.

Window and Roof Ventilation   When the roof or some roof features must be opened for venting, the windows on the top floor should be opened after the roof is opened. If this cannot be done from inside, the windows can be broken with tools operated from the roof or from a ladder, fire escape, porch, or balcony **Figure 4-5**.

When the windows on several stories must be opened, fire fighters should begin at the top and work down. Fire fighters going up to the top floor on a fire escape, for example, should not open windows as they go because this would allow smoke to flow out of the windows and possibly engulf them. Flames also might issue from the windows and trap the fire fighters.

If the fire or trapped gases should blow out a window above the fire fighters who are working below, they should seek shelter in doorways or away from the building to avoid flying glass and other debris. They should never look up to see what is happening. Instead, they should take cover until the danger is past.

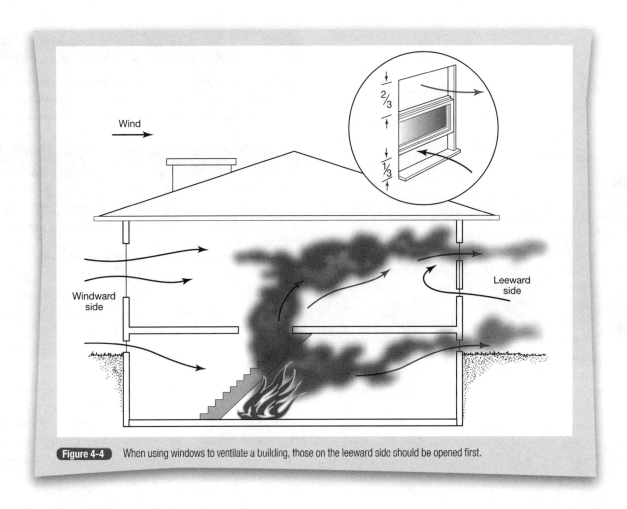

Figure 4-4   When using windows to ventilate a building, those on the leeward side should be opened first.

Natural Roof Openings

In multistory buildings, vertical shafts carry stairways, elevators, dumbwaiters, electric wiring, heating and air conditioning ducts, plumbing, and waste pipes. These shafts usually extend through the full height of the building. Convected heat, smoke, and gases may rise within and around them when the building is involved in fire. If this occurs and the shafts are not opened at the roof, fire will travel horizontally from the tops of the shafts, possibly igniting the top of the structure, and then work its way down.

The slight pressure that will develop at the top of the shaft will force heat, smoke, and gases throughout the upper parts of the building, pushing the products of combustion into other shafts, into hallways, and into apartments, offices, or other areas.

> ### Key Points
>
> Convected heat, smoke, and gases may rise within and around shafts when the building is involved in fire. If this occurs and the shafts are not opened at the roof, fire will travel horizontally from the tops of the shafts, possibly igniting the top of the structure, and then work its way down.

The greater the buildup of combustion products, the greater is the probability of fire on the upper floors. Shafts are usually capped at the roof with various types of closures. These coverings can be removed to provide effective openings into the building.

Skylights

The positioning of skylights can give fire fighters an idea of the layout of the building under them. For instance, in an apartment or office building, a row of skylights from front to rear is most likely located over the top-floor corridor. In shops or factories, a line of skylights is usually placed over a work area. In apartment buildings, office buildings, and similar structures, individual skylights are often located over stairways, corridors, elevator shafts, air shafts, and bathrooms. Those placed over bathrooms usually have louvered ends to allow normal venting.

The area immediately below a skylight usually is boxed in so the cockloft or attic space is effectively separated from the skylight. Thus, when the skylight is opened, the building proper will be ventilated, but the space just below the roof will not. To ventilate this space, the roof or the boxed-in area must be opened.

Fire fighters may be working in the area below a skylight; therefore, before the skylight is removed or its glass removed,

Figure 4-5 Following roof ventilation, the top floor windows should be opened.

Key Points

The area immediately below a skylight usually is boxed in so the cockloft or attic space is effectively separated from the skylight. Thus, when the skylight is opened, the building proper will be ventilated, but the space just below the roof will not. To ventilate this space, the roof or the boxed-in area must be opened.

Key Points

Fire fighters may be working in the area below a skylight; therefore, before the skylight is removed or its glass removed, some warning must be given to those below by banging on the roof at the base of the skylight or on the sides of the skylight at the roof line and via the radio.

some warning should be given to those below by banging on the roof at the base of the skylight or on the sides of the skylight at the roof line. Fire fighters working below the skylight must be aware of the meaning of this signal and should take cover. The signal should be given even if ladder crews intend only to remove or tip the skylight, as the glass might accidentally break, plastic may break in pieces, or part of the skylight might fall in during the process. Portable radios are an excellent way to communicate with fire fighters working below the area and those on the roof.

Opening Skylights There are three ways to open a skylight for venting. The preferred method is to either lift the skylight from its opening or tip it over onto the roof. The flashing that joins the skylight to the roof must first be cut or pried away. If this can be done quickly on all four sides, the skylight can be lifted off the opening, as shown in **Figure 4-6a**. Otherwise, the flashing should be removed from the two short sides and from one long side and the skylight and then tipped over using the fourth side as a pivot, as shown in **Figure 4-6b**. The glass in the skylight can also be removed in one piece or broken, as shown in **Figure 4-6c**.

Some skylights are not mounted directly on the roof, but are instead placed on a wooden foundation approximately 6 inches above the roof. The flashing is then attached to the wooden foundation rather than to the roof. In this case, too, the flashing must be cut or pried loose before the skylight can be removed.

To ensure that fire fighters do not fall through the opening where a skylight has been lifted away, the skylight should be laid on the roof upside down, as shown in Figure 4-6a to serve as a warning. Then it will not be mistaken in smoke or darkness as being in place over its opening.

If for some reason the skylight cannot be lifted or tipped, the glass or plastic covering can be removed. In glass applications, the least damage is done by peeling back the metal stripping

Key Points

To ensure that fire fighters do not fall through the opening where a skylight has been lifted away, the skylight should be laid on the roof upside down to serve as a warning.

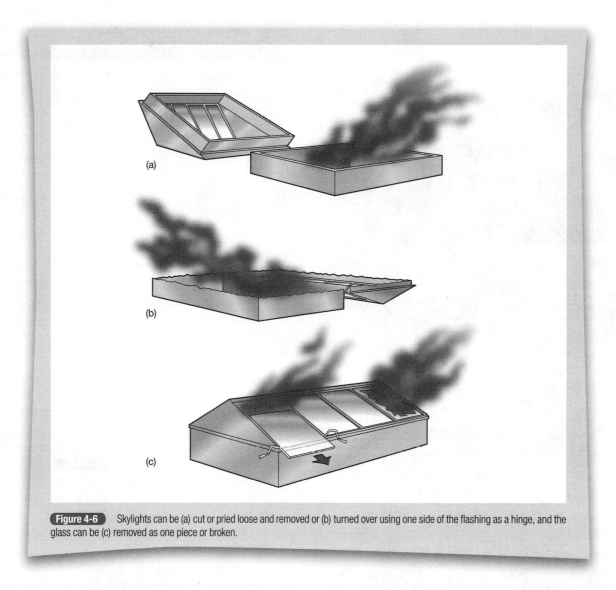

Figure 4-6 Skylights can be (a) cut or pried loose and removed or (b) turned over using one side of the flashing as a hinge, and the glass can be (c) removed as one piece or broken.

along the bottom edge of the glass and sliding the panes out, as shown on Figure 4-6c. If the glass will not come out easily and quickly, it must be knocked out. Although this method is the least desirable, it might be the only way to quickly open the skylight for venting. The skylights installed over garages, stores, shops, factories, and other large open areas can be large and very heavy. One or two fire fighters may not be able to lift or tip this type of skylight and will have to remove or break the glass.

Plastic Skylights and Roof Panels Various sizes of clear plastic "bubble-type" skylights are installed in some structures. These are made by setting a one-piece unbreakable plastic bubble in a frame. Plastic skylights are mounted in the same way as glass skylights and can be lifted or tipped in much the same manner. When a plastic skylight cannot be removed quickly, the frame should be cut where it meets the plastic and the bubble pried up **Figure 4-7** .

Clear, frosted, or colored plastic panels are sometimes placed in a roof to serve as a simple skylight. Because these panels are usually weak and will not support much weight, ladder crews should avoid stepping on them. These panels are used mainly in the gable roofs of modern noncombustible buildings, such

Key Points

If for some reason the skylight cannot be lifted or tipped, the glass or plastic covering can be removed.

Key Points

Clear, frosted, or colored plastic panels are sometimes placed in a roof to serve as a simple skylight. Because these panels are usually weak and will not support much weight, ladder crews should avoid stepping on them.

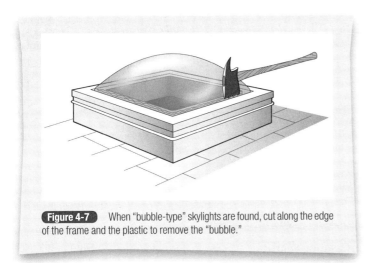

Figure 4-7   When "bubble-type" skylights are found, cut along the edge of the frame and the plastic to remove the "bubble."

as warehouses, factories, garages, and shops. The roof itself is often constructed of corrugated metal with no ceiling below. The plastic panels can be pulled up after the roof is cut or pried up along one edge of the panels **Figure 4-8** .

Fire fighters working on the roof should be made aware of any openings that would cause serious injury or death if someone fell through.

Effects of Wind   Any large opening in the roof, such as a skylight opening, will allow fire, heat, and smoke to rise through the roof. For this reason, ladder crews opening a roof or any roof feature should keep their backs or sides to the wind and wear full PPE, including SCBA. If they must face the wind, they will be protected from a blast of fire, heat, or smoke and gases.

Key Points

Any large opening in the roof, such as a skylight opening, will allow fire, heat, and smoke to rise up through the roof. For this reason, ladder crews opening a roof or any roof feature should keep their backs or sides to the wind and wear full PPE, including SCBA.

Figure 4-8   Plastic panels are sometimes placed in the roof to serve as skylights.

Openings Below the Skylight   After the skylight is removed or knocked out, smoke should flow freely from the opening. If it does not, there might be a swinging transom or a glass panel at the ceiling line below the skylight. The transom should be opened if possible or its glass removed. If this can be done quickly, it should be broken. This should be done carefully to avoid the heat and smoke that will flow out of the opening.

If the area below the opened skylight is boxed in, the cockloft must be vented through an opening in the roof. A roof scuttle (discussed later) can be used for this purpose, if it is located in the right spot and is not boxed in like the skylight. If no such roof feature is available, it will be necessary to open the boxed-in area below the skylight or the roof cut open around the skylight **Figure 4-9** .

If the fire is directly under the roof (that is, on the top floor of the building), a roof opening should be made as close over the fire as safety allows. Otherwise, the opening will draw the fire across the top of the fire floor, under the roof or ceiling.

When the roof is separated from the fire by at least one story, the roof can be opened around the skylight after the skylight is opened. In this case, the boxed-in area can be opened instead of the roof, but if smoke and heat issuing from the skylight shaft make this impossible, then the roof will have to be opened.

Figure 4-9   To vent the cockloft or attic, the roof or the closed-in area beneath the skylight must be opened. Removing the skylight alone will not vent the area under the roof in such cases.

Roof Scuttles (Hatches)

A **scuttle** is placed in a roof to allow access to the roof from inside the building. Sometimes a ladder is built into the wall below the scuttle. In multiple occupancies, scuttles are usually located above and at one end of the top-floor corridors. In stores and other business establishments, they are usually located at the rear of the building.

Scuttles vary in size but are usually square in shape **Figure 4-10**. A scuttle consists of a metal or wooden cover that fits tightly over a raised support attached to the roof. The wooden cover is encased in roofing metal and is sometimes tarred. It

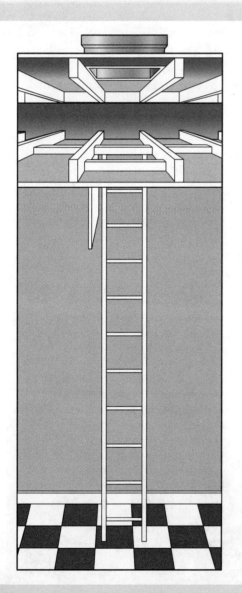

Figure 4-10 Opening scuttles can be an aid to venting. If the scuttle is locked securely, cut out the top. After open, check for an enclosure through the attic space and for an opening at ceiling level.

is often held in place only by the support. Over the years, rust, corrosion, grit, and dirt that get between the cover and support can tighten the seal. To prevent burglaries, the covers are sometimes bolted or padlocked in place on the inside; steel bars might also secure them.

Opening Scuttles A cover that is not securely locked in place can be pried off with a pick axe, Halligan bar, claw tool, or similar device. If a cover cannot be removed quickly, the top of the cover should be cut out. It can easily be replaced.

Opening Below the Scuttle As with skylights, the area below a scuttle might be boxed in to separate it from the cockloft. After the cover is removed or cut, ladder crews should check to see if the scuttle is closed at ceiling level. If smoke pours out, these areas will have to be probed with tools.

If the scuttle is open at the ceiling level but boxed in through the cockloft, the building is being vented. If smoke is coming out but there is a panel at the ceiling level, the cockloft is being vented. If there is no closure at the ceiling level or the cockloft, both the building and the cockloft are being vented.

If there is a closure at the ceiling level, it must be removed or knocked out. This should be done with an axe, a ceiling hook, or some other tool, depending on the depth of the opening. If the enclosure below the scuttle is boxed in, the cockloft is not being ventilated. Such enclosures are usually made of very light tin, thin wood, or plaster that can be removed quickly; some, however, are made of a tougher material that is difficult to open.

Key Points

Ladder company personnel should be familiar with the different types of ventilators and the type of venting each provides.

Ventilators

There are various types, sizes, and shapes of ventilators and vent pipes, and each usually serves a different purpose. A ventilator customarily caps the vent pipe. Some ventilators open only into the cockloft. Others open into the top floor of the building; these are used to exchange air and equalize pressures within the building. Others may be on vertical shafts that extend the full height of the building; these usually connect into heat-producing areas, such as kitchens and laundries, and might contain fans to increase the flow of air. Still others are used to ventilate bathrooms in multiple occupancies, including apartments, office buildings, hotels, and motels; these usually contain blowers or fans. The huge ventilators on the roofs of theaters and auditoriums usually open into the main assembly area at ceiling level; their ceiling inlets are normally covered with decorative interior grilles.

Each of these ventilators can be used to vent a building. Ladder company personnel should be familiar with the different types of ventilators and the type of venting each provides **Figure 4-11** .

A louvered ventilator that can be rotated by the wind or by rising heat, as shown in **Figure 4-12a** is usually connected to a shaft that runs the full height of a multistory building. In a single-story building, this type of ventilator is usually placed directly over a heat-producing area. In a residence, it is used to vent the attic space to assist in cooling the structure.

In an apartment house, ventilators that cap the fan shafts leading from kitchens or bathrooms are normally lower than other ventilators, and they have a different appearance. They often curve as they come out of the roof so that they point down toward the roof, as shown in **Figure 4-12b** . Some are shaped like small skylights; they are made of roofing tin with a solid top, as shown in **Figure 4-12c** . Their open ends are covered with hardware cloth or wire to permit the venting action.

Other ventilators, the stationary type, are not connected to heat-producing areas. These consist of a pipe extending above the roof, a cover located an inch or two above the pipe, and a

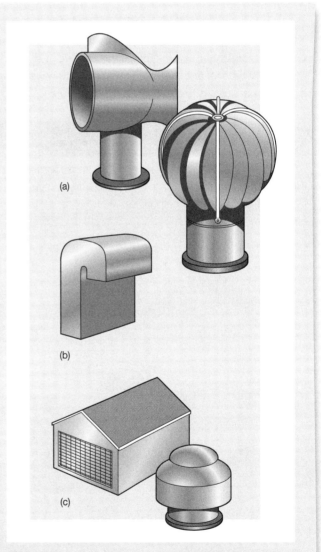

Figure 4-12   There are various types of ventilators: (a) louvered ventilators rotated by wind or rising heat are typically connected to a shaft on multistory buildings or above heat producing areas in single-story buildings and attic spaces in residences; (b) ventilators that cap fan shafts in an apartment house often curve as they come out of the roof; (c) ventilators with open ends are covered with cloth or wire to permit the venting action.

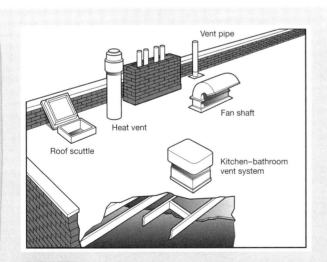

Figure 4-11   Ventilators can aid ladder crews in venting a building, and fire fighters should recognize the different types.

wide metal band that passes around the top of the pipe and the cover. The cover and band keep rain, snow, and dirt from being blown into the pipe.

Removing Ventilators Smoke coming from a ventilator indicates that the fire has reached the area it serves. The ventilator should be opened to remove the restriction at the top (the weather cover), as shown in **Figure 4-13**.

A stationary ventilator can be removed by getting a hold on the underside of the metal band with the pick of an axe, a Halligan bar, a claw-tool hook, or similar tool and pulling off the band. The cover usually comes off with the band; if not, it can be pulled away from the pipe. The amount of smoke leaving the ventilator may increase dramatically after the cover is removed.

Rotating ventilators are usually stronger than the stationary type. If a rotating ventilator is difficult to remove, it should be broken off at the roof line. Although this is not a desirable action, as it leaves a hole in the roof, it may be necessary. The low

Figure 4-13 Smoke issuing from a ventilator is an indication that it should be opened for full effect.

Key Points

Smoke coming from a ventilator indicates that the fire has reached the area it serves. The ventilator should be opened to remove the restriction at the top (the weather cover).

ventilators over fan shafts can usually be pulled up from the shafts. The curved type should be pulled off at the roof line, as shown in **Figure 4-14**.

Sheet Metal Shafts It is especially important that ladder crews be able to recognize ventilators that cover sheet-metal shafts.

Figure 4-14 Ventilators over fan shafts should be removed to assist in venting the building. Fans should be shut down.

Because the metal is usually very thin, the heat of a fire traveling through such a shaft can ignite combustible materials in contact with the shaft. In addition, the fire might extend through seams in the shaft and ignite building features around it. This is especially true of grease duct vents in restaurant and hotel kitchens.

The roof area around this type of shaft should be checked for heat if the building is involved with fire. If the roof is hot and the tar shiny or if the tar is dry on a wet surface, the roof should be opened around the shaft or its ventilator. This will allow fire around the shaft to be drawn up and out of the building and will deter lateral fire spread.

Plumbing System Vent Pipes   Each separate plumbing system in a building has a vent pipe that extends through the roof. These pipes are about 2 inches in diameter, and thus, they can be run vertically through walls or partitions. Because they pass through each floor of a multistory building, they must be considered as vertical shafts and as possible channels for the spread of fire and combustion products.

The roof area around each plumbing system vent pipe should be checked. If it shows signs of heat or smoke, the roof should be opened around the pipe.

Penthouses (Stairway Covers; Bulkhead Doors)   A **penthouse** is a small hut-like enclosure built over a stairwell that allows the stairs to extend up to the roof level and is tall enough to permit a person to walk onto the roof. The penthouse has a full-sized door at the roof level, and there can also be a door at the top floor. These doors can be kept locked or unlocked. The penthouse itself might have windows in its sides and a skylight on its roof.

Because a penthouse covers an open stairwell, it can be used to ventilate the entire height of the stairwell. Opening the penthouse will also ventilate corridors and hallways that are open to the stairs, as shown in **Figure 4-15**. Because the stairs and corridors will be used by occupants attempting to leave the fire building and by engine crews advancing their attack lines, penthouses should be opened as soon as possible. If more than one is available or

Figure 4-15   Opening the penthouse can aid in ventilation of stairway and corridors.

if stairs lead directly to the roof, one stairway could be used for ventilation while the other is used for egress from the building or by fire fighters to gain access to upper floors.

Heavy smoke showing around a penthouse usually indicates that it is open to the stairway. The roof door is opened in the same way as any other door. If it is locked and a key is not available, the door should be opened by force. If there is a closed door at the top floor level, it must be opened immediately.

Before the lower door is opened, it should be checked for signs of heat. If the door is very hot, a blast of heat might hit the fire fighters when they open it, and smoke might obscure their vision. They may have difficulty getting back up the stairs to the roof. Fire fighters should use extreme caution in opening a door to check the stairwell if fire or products of combustion are suspect.

Machinery Covers

Small, box-like enclosures are sometimes found on the roofs of office buildings, apartments, and commercial buildings. These machinery covers usually house dumbwaiter pulleys or motor assemblies, the valves for heating and air conditioning systems, and similar items. They are usually made of wood and covered with roofing metal; sometimes they are tarred. They can also be equipped with ventilators.

Key Points

When fire fighters detect smoke issuing from a machinery cover or heat in its walls, they should vent it. This can usually be done simply by opening its service door if so equipped.

When fire fighters detect smoke issuing from a machinery cover or heat in its walls, they should vent it. This can usually be done simply by opening its service door if so equipped.

Elevator Penthouses

If there is an elevator in a building, there is probably an elevator penthouse on the roof. The elevator penthouse contains the motors and electric switches that control the elevators. It is located directly above the elevator shaft and is open to the shaft.

By ventilating the elevator penthouse, fire fighters can reduce the accumulation of heat and smoke in the shaft. At the same time, removing excessive heat from around the motors and switches in the house lessens the chance of machine damage, which could result in erratic operation of the elevators. Even if there is fire in the elevator shaft and the elevators are not being used, the elevator penthouse should be ventilated to help draw the fire out of the building and deter lateral fire spread. The use of elevators during a fire in a building must be made by command based on department standard operating guidelines and a risk-versus-benefit analysis.

Elevator penthouses vary in size depending on the number of elevators in the building. They range from small structures no bigger than an elevator car to huge structures serving several elevators. The larger penthouses often have windows that can be opened or broken for ventilation purposes. The door to the elevator penthouse also could be opened by force if necessary to provide ventilation. If there is a skylight on the roof of an elevator penthouse, it, too, can be opened for ventilation; however, these skylights are often difficult to reach without a short ladder. It is

Key Points

Even if there is fire in the elevator shaft and the elevators are not being used, the elevator penthouse should be ventilated to help draw the fire out of the building and deter lateral fire spread.

Key Points

Working in an elevator penthouse can be extremely dangerous. Electrical circuits, motors, and pulleys present a hazard to fire fighters even when there is not a fire. Use extreme caution in this area if the elevators are to remain in operation.

Key Points

Most air shafts are not covered at the roof, and thus, the inner rooms of the building can be ventilated by opening these windows.

usually quicker to vent through the door. Working in an elevator penthouse can be extremely dangerous. Electrical circuits, motors, and pulleys present a hazard to fire fighters even when there is not a fire. Use extreme caution in this area if the elevators are to remain in operation.

Air Shafts

Air shafts are usually found in older buildings. Air shafts might be located entirely within one building or between two row structures. Their purpose is to allow light and air to enter the inner rooms. Windows in these rooms open into the air shaft at each floor. Most air shafts are not covered at the roof, and thus, the inner rooms of the building can be ventilated by opening these windows.

Some four-sided shafts have skylight-type coverings at the roof, as shown in **Figure 4-16**. These are usually referred to as **light shafts.** If fire has reached such a shaft, the skylight should be removed or opened so that heat will not build up in the shaft and be forced into the rooms around it. The fire should be reported immediately so that hose lines can be placed to cut off its spread. The skylight should also be opened if the shaft is being used to vent inner rooms.

If the skylight cannot be lifted or tipped quickly, the glass panes should be removed or broken. The shaft will not be used by fire fighters or escaping occupants, and thus, there is little danger of glass falling on anyone.

Preincident Planning

Although few, if any, buildings will contain all the natural openings discussed in this section, every building may have some of them. The only sure way to know which natural openings can be used in fighting a fire is by making preincident inspections.

Key Points

If fire has reached a light shaft, the skylight should be removed or opened so heat will not build up in the shaft and be forced into the rooms around it.

Key Points

The only sure way to know which natural openings can be used in fighting a fire is by making preincident inspections.

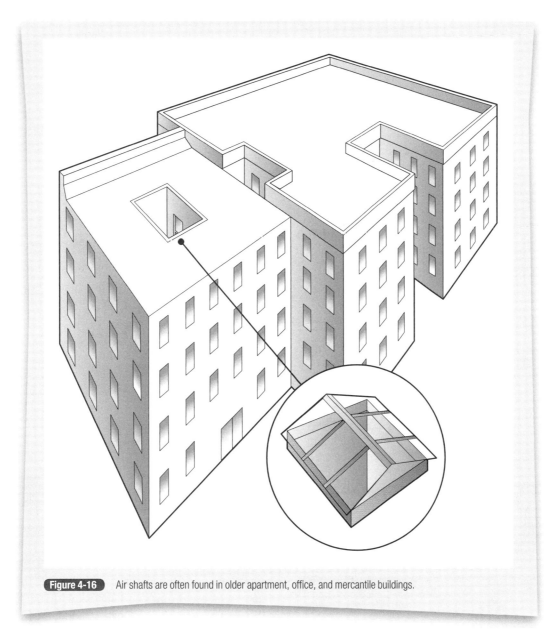

Figure 4-16 Air shafts are often found in older apartment, office, and mercantile buildings.

A check through one apartment building in a housing project will usually enable ladder crews to determine which roof features vent a particular area or building component. For instance, a visual check for vertical story-to-story rows of small frosted windows will indicate the locations of bathrooms and the accompanying pipe shafts that rise up through the building. The information gained in one building will apply to all similar buildings in the project. Preincident planning can save time and effort on the fire ground; fire fighters become familiar with the building and that familiarity allows them to make pertinent decisions if a fire occurs.

Cutting Through Roofs

At times, the only way to properly ventilate part or all of a building is to cut a hole in the roof. A roof made of boards under standard roofing materials can be cut with axes and the cut areas forced open with Halligan bars, claw tools, or the pick of an axe. A plywood roof should be cut with a power saw, if available, because it is difficult and time consuming to cut through plywood by hand. In either case, fire fighters should be careful not to cut joists or other structural members; this will weaken the area in which they are working **Figure 4-17**.

A single large hole is more effective than several small holes and is also safer for fire fighters operating on the roof. One

Figure 4-17   If it is necessary to open a roof, a large hole should be made. Care should be taken to avoid cutting structural members.

Key Points

A single large hole is more effective than several small holes and is also safer for fire fighters operating on the roof.

4- × 8-foot hole has twice the area of four 2- × 2-foot holes and requires less cutting.

All of the roof boards should be cut through before any of them are pulled up. If a board is pulled up as soon as it is cut, the

Key Points

All of the roof boards should be cut through before any of them are pulled up. If a board is pulled up as soon as it is cut, the smoke pouring through the small hole could drive fire fighters from their position, making it impossible to complete ventilation.

smoke pouring through the small hole could drive fire fighters from their position, making it impossible to complete ventilation. When pulling up the cut boards, ladder crews should always keep their backs to the wind.

When all the boards have been cut, the ceiling below should be knocked down with a pike pole or similar tool. The ceiling hole should be made as large as the roof hole. There is little sense in making a roof hole of the proper size and then restricting ventilation with a tiny ceiling hole.

Special care must be taken when the fire is immediately below the roof; that is, in the attic under a gabled roof or in the

Key Points

The ceiling hole should be made as large as the roof hole. There is little sense in making a roof hole of the proper size and then restricting ventilation with a tiny ceiling hole.

cockloft under a flat roof. In such cases, the roof must be opened as close as possible to the seat of the fire. Otherwise, the fire, along with heat and combustion products, will be drawn across the top of the building to the opening. Heat rising from the fire will develop a hot spot on the roof that will indicate the location of the greatest concentration of fire.

If the roof is flat, the opening should be made at the hot spot or as close to it as fire fighters can get safely **Figure 4-18**. If the roof is pitched and the hot spot is at the peak, the opening should again be made at the hot spot. Otherwise, a pitched roof should be opened so that the hole extends from the hot spot toward the peak of the roof to speed up the venting of accumulated smoke and gases **Figure 4-19**. The draft can be increased by making an additional opening just above the eave line on the side of the roof opposite the original opening.

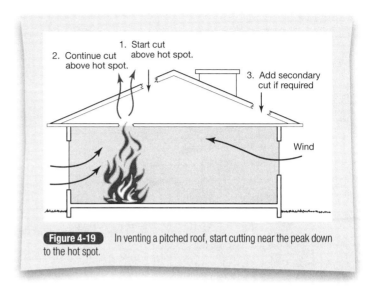

Figure 4-19 In venting a pitched roof, start cutting near the peak down to the hot spot.

Figure 4-18 On a flat roof, the opening should be made as close to the hot spot as safety allows.

The same sort of care must be exercised when a natural roof opening is used to vent an attic or top-floor fire. By removing a scuttle or skylight that is some distance from the fire, ladder crews might actually increase fire spread. If the hot spot is not near enough to a natural opening, the roof should be cut.

Fire fighters working on any type of roof including flat, pitched, or arched should use extreme caution. Fire fighters should be familiar with the various types of roof construction and designs in their area. Fire fighters must be in complete personal protection equipment (PPE), including SCBA, and be in communication with their sector officer if working within a sector or command. Roof operations should be coordinated with companies working below. When working on a roof, fire fighters must be aware of the location and extent of the fire. Members should sound the roof and be aware of tell-tale signs of unsafe roof conditions indicating where the fire may be. Only those required to perform the required roof operation should be there. Always provide a secondary means of escape off the roof. Fire fighters should use roof ladders and safety or rescue rope when working on pitched roofs, as shown in **Figure 4-20**. Use extreme caution when using hand or power tools, ensuring that no one will come in contact with the equipment. When the task is completed, fire fighters should leave the area. Working on a roof under fire conditions is not an easy task. Fire fighters should not lose track of the environment they are working in and should think of safety as their number one priority.

Key Points

Fire fighters working on any type of roof including flat, pitched, or arched should use extreme caution. Fire fighters should be familiar with the various types of roof construction and designs in their area.

Figure 4-20   Fire fighters should use a roof ladder and safety rope when venting a pitched roof.

Forced Ventilation

Forced, or mechanical, ventilation is usually accomplished through the use of smoke ejectors or blowers or hydraulically with fog streams. Forced ventilation may help the task move along more quickly than with just the use of natural openings. Smoke ejectors can be operated electrically, by gasoline-driven engines, or hydraulically by water pressure from hose lines. A fog stream is usually used to hasten the removal of smoke from an area after the stream has extinguished the fire.

Smoke Ejectors

Smoke ejectors could be used as an additional method for the venting techniques described earlier in this chapter. The heat, smoke, and gases produced by a working fire are most effectively removed by natural convection through natural or forced openings. Ladder crews usually perform ventilation activities

Key Points

Forced ventilation may help the task move along more quickly than with just the use of natural openings.

Key Points

Smoke ejectors could be used as an additional method for the venting techniques described earlier in this chapter. The heat, smoke, and gases produced by a working fire are most effectively removed by natural convection through natural or forced openings.

using hand tools. A sizeup of conditions will usually dictate when forced ventilation is needed.

There are both advantages and disadvantages to using forced ventilation. The major advantage is that it speeds the removal of the contaminated area removing smoke, heat, and toxic gases. Introducing air under pressure could cause the fire to intensify and spread, especially in areas where fire fighters may have not made entry. The use of forced ventilation must be supervised to ensure that it is not misapplied and properly controlled. A disadvantage is that smoke ejectors must have a power source to operate.

Limitations   Smoke ejectors should not be used in partially or completely confined spaces, such as attics, corridors, or closed-up basements, in which there is a working fire. Under these conditions, the smoke ejectors could spread the fire laterally. They can be used in confined spaces after the fire has been knocked down, but even then, fire fighters operating hose lines must keep close watch to ensure that the air movement does not fan embers into open flame.

In spite of such problems, smoke ejectors can be of great help on the fire ground. In some situations, discussed in the next chapter, they can constitute the major tool in a venting operation. In others, they might be used to supplement more standard ventilation techniques.

Smoke Ejector Placement   Smoke ejectors are traditionally used to provide negative pressure ventilation. Negative pressure uses fans to pull smoke out of a building. Fans are most effective when placed where they tend to increase natural airflow. They should, therefore, be positioned in windows, doorways, roof openings, basement openings, or openings that have been made to ventilate the building. The ejector should be positioned in a window or other opening to exhaust in the same direction as the natural wind current. Ejectors in windows are usually

Key Points

Smoke ejectors should not be used in partially or completely confined spaces, such as attics, corridors, or closed-up basements, in which there is a working fire. Under these conditions, the smoke ejectors could spread the fire laterally.

Key Points

Smoke ejectors are most effective when placed where they tend to increase natural airflow. They should, therefore, be positioned in windows, doorways, roof openings, basement openings, or openings that have been made to ventilate the building.

Key Points

Positive-pressure ventilation is a technique that uses a forced airflow to create pressure differentials. This method uses a high-volume fan, much larger than the traditional smoke ejector, to create a higher pressure inside the building than that of the outside atmosphere.

placed high because that is where the products of combustion have collected.

Smoke ejectors can be effective especially when used in pairs. An additional method that can be employed is to place another fan at an opening opposite the other. This ejector will be used to blow fresh air toward the other fan and should hasten the removal of the products of combustion.

When a smoke ejector is positioned in a window or doorway, all shades, drapes, blinds, curtains, and screens should be removed to eliminate restriction of airflow. If possible, the open area around a smoke ejector should be closed with salvage covers or whatever materials are available; this increases its efficiency by directing air to and through the opening and by preventing smoke from reentering **Figure 4-21** .

Keep the flow of air in a straight path if possible and remove any obstacles that may hinder the flow of air to the ejector. Fire fighters must be careful when positioning a smoke ejector not to exhaust smoke into congested areas. It must be placed so that no smoke is blown into the open windows of nearby buildings or into the intakes of heating and cooling systems.

Positive-Pressure Ventilation

Positive-pressure ventilation is a technique that uses a forced airflow to create pressure differentials. This method uses a high-volume fan, much larger than the traditional smoke ejector, to create a higher pressure inside the building than that of the outside atmosphere. With pressure greater inside the building,

the products of combustion seek an outlet to the outside through openings created by fire fighters.

In most cases, the fan is set up outside an exterior doorway. The cone of the airflow should cover the opening in the doorway. On the opposite side of the building, an exit opening, usually a window, is created. This opening should be as large, if not somewhat larger, than the entry opening. This should be the only exit opening created. Fire fighters inside the building can control the flow of air by opening and closing doors to facilitate the operation.

Over the years, much discussion has taken place as to the pros and cons of using positive-pressure ventilation. Some fire departments are true advocates of its use, whereas others are not comfortable with its disadvantages.

Among the advantages of positive-pressure ventilation is the ability to set the fan up outside the building without entering the structure. If applied properly, it will ventilate an area more efficiently than negative-pressure ventilation. It can be used in large buildings or areas and will work both for horizontal and vertical ventilation. With the fan in operation, fire fighters can enter an area in front of the fan, which is blowing fresh air into the structure.

The disadvantages include the coordination of the operation. When do the fans get set up, and who makes the exit opening? The incident commander must have a strong presence to

Figure 4-21 For greatest effectiveness, the open area around the smoke ejector should be closed.

Key Points

In most cases, the fan is set up outside an exterior doorway. The cone of the airflow should cover the opening in the doorway. On the opposite side of the building, an exit opening, usually a window, is created. This opening should be as large, if not somewhat larger, than the entry opening.

Key Points

Over the years, much discussion has taken place as to the pros and cons of using positive-pressure ventilation. Some fire departments are true advocates of its use, whereas others are not comfortable with its disadvantages.

Key Points

Hydraulic ventilation is typically used to clear an area of smoke, heat, and gases after the fire has been knocked down.

coordinate this task. If horizontal or vertical ventilation has taken place before hand, there may be many openings throughout the structure.

There may be a concern of performing other ladder company tasks initially with the personnel assigned on the first alarm. There is also a belief that by using positive-pressure ventilation the fire may intensify, particularly in areas that may be hidden, such as wall and ceiling voids, and in areas where fire fighters are not working.

Hydraulic Ventilation

Hydraulic ventilation is typically used to clear an area of smoke, heat, and gases after the fire has been knocked down.

A spray nozzle set on a fog pattern can be used to start ventilating immediately after the fire has been knocked down in a room. To perform this operation, the stream must be directed out of a window in such a way that it draws out the remaining smoke and gases.

To be most effective, the stream should be positioned so the fog pattern covers approximately 90% of the window opening. This usually means that the nozzle should be held a few feet back inside the window. A good way to determine where to

Key Points

To be most effective, the stream should be positioned so the fog pattern covers approximately 90% of the window opening.

position the nozzle is to start by holding the open nozzle outside the window and then slowly backing it into the room while observing the movement of the smoke. At the proper nozzle position, the stream will draw the smoke out the window in large volumes. The size of the opening will determine rate of smoke discharge from the window. The larger the window is the faster the ventilation will be accomplished.

Fog streams should not be used for venting if they will damage items removed to the outside for protection, if the streams enter adjoining buildings or areas where they may cause additional damage, if they will do unnecessary damage inside the room from which they are directed, or if they will cause ice to form outside, thus creating a hazard to fire fighters.

Fog stream venting should be a short-term operation. If the operation is obviously ineffective in a particular situation, other venting techniques should be used. Both the personnel and the hose lines can probably be used to better advantage, especially if the fire force is short in numbers.

Solid streams, although less effective, can be used in a similar manner, either by opening the nozzle halfway or by removing the tips and using the shutoff opened fully. The broken stream created will move air to clear the area.

Chief Concepts

- Ventilation is a primary concern of fire fighters working on the fire ground. It allows fire fighters to get inside the building and begin a primary search. With tenable conditions within the building, fire fighters are able to perform search and rescue operation and locate and extinguish the fire in a timely manner. Ventilation involves opening a fire building to the outside so that heat, smoke, and gases can exit, mainly by natural convection. In many cases, fire fighters can make use of natural openings, thereby minimizing the damage done during ventilation operations; however, when required by the fire situation, there should be no hesitation in opening the roof and top floor of a building to ventilate it. Savings in terms of deaths, injuries, and property damage far outweigh the cost of repairing the building.

- Among the roof features that can be opened to ventilate all or part of a building are vent pipes with ventilators, roof scuttles, skylights, machinery covers, elevator penthouses, and stairway penthouses. In addition, forced ventilation using fans and smoke ejectors and hydraulic ventilation using fog streams can be used to enhance ventilation.

- Ventilation is usually performed by ladder company members who must be properly trained, adequately staffed, and supported with the proper tools and equipment. Ventilation operations must be a coordinated effort between ladder company and engine company members. Charged hose lines should be in place before, during, and after ventilation operations take place. Fire fighters should be cognizant of the fact that fire could spread throughout the building if improper ventilation practices are employed.

Key Terms

Light shafts: Four-sided shafts with skylight-type coverings.

Mushrooming: The shape of combustion products on the fire floor when they have accumulated under the roof and are forced to move horizontally.

Penthouse: A small hut-like enclosure built over a stairwell that allows the stairs to extend up to the roof level and is tall enough to permit a person to walk onto the roof.

Scuttle: Construction feature of a roof that allows access to the roof from inside the building.

1. Hot air, smoke, and heated gases traveling through a building via _____ heating present the greatest firefighting problem.
   a. Induction
   b. Conduction
   c. Radiant
   d. Convection

2. Smoke and heated gases within a structure will normally
   a. Travel upward and then move horizontally if there is an upward obstruction and finally downward
   b. Travel horizontally until meeting an obstruction then upward through the building, usually via concealed spaces
   c. Travel in all directions simultaneously
   d. The movement of smoke and heated gases is nearly impossible to predict under fire conditions.

3. This is the general rule behind all ventilation operations: Open the fire building in such a way that all accumulations of heat and the products of combustion will leave the building by natural _____.
   a. Induction
   b. Conduction
   c. Convection
   d. Radiation

4. Roof ventilation is rarely needed in one- and two-story dwellings unless
   a. The fire has entered the walls.
   b. The fire is moving upward via the walls or stairways.
   c. The fire is in the attic.
   d. Roof ventilation should always be performed when there is a working fire in a structure.

5. When venting double-hung windows
   a. Completely remove the glass and window frame.
   b. Open the bottom or break out the bottom pane only.
   c. Open the top or break out the top pane only.
   d. Open two thirds from the top and one third from the bottom.

6. When wind is a factor and windows are to be used for ventilation
   a. The windows on the leeward side should be opened first followed by opening the windows on the windward side.
   b. The windows on the windward side should be opened first followed by opening the windows on the leeward side.

   c. Only the windows on the leeward side should be opened.
   d. Only the windows on the windward side should be opened.

7. When the roof is opened for ventilation
   a. The windows on all floors should be opened immediately after the roof is opened.
   b. The windows on the top floor should be opened after the roof is opened.
   c. The windows in the building should remain closed if possible.
   d. The windows on the fire floor should be opened after roof ventilation is complete.

8. When windows on several stories must be opened via a fire escape or other exterior access, fire fighters should
   a. Open the windows in the most expeditious manner.
   b. Begin at the fire floor and work upward.
   c. Begin at the top floor and work downward.
   d. Begin at the lowest floor and work upward.

9. In a factory building, rows of skylights are normally located over
   a. Work areas
   b. Large open stairways
   c. Bathrooms
   d. Corridors

10. The *least* desirable way to use a skylight for ventilation is by
    a. Entirely removing the skylight by lifting it off its opening
    b. Removing three sides and tipping the skylight by using the fourth side as a hinge
    c. Breaking the glass or plastic in the skylight
    d. Any of the above is equally safe and effective.

11. To ensure that fire fighters do not fall through the opening created when a skylight is lifted away
    a. The skylight should be placed back over the opening as soon as the first heavy volume of smoke is released.
    b. The skylight should be placed back over the opening at a right angle to the original placement.
    c. Lay the skylight upside down on the roof near the opening.
    d. Skylights should not be lifted off the opening; it is better to either remove the glass/plastic or use the tip method.

12. When opening the roof for ventilation, fire fighters should

    a. Keep their back to the wind.

    b. Keep their backs or side to the wind and wear full PPE and SCBA.

    c. Wear full PPE and SCBA and approach the fire from the side they entered regardless of wind conditions.

    d. Keep their face or side to the wind and wear full PPE and SCBA.

13. It is best to cut a plywood roof using a

    a. Power saw

    b. Chisel head axe

    c. Flat head axe

    d. Pick axe

14. A large single ventilation hole in the roof is preferred over several smaller holes because

    a. It is more effective.

    b. Is safer for fire fighters.

    c. It requires less cutting.

    d. All of the above.

15. When cutting roof boards for ventilation

    a. Pull up each board as it is cut.

    b. Wait until all boards are cut before pulling up any roof boards.

    c. Pull up all boards on one side of the cut before proceeding to the other side to complete the cut.

    d. Pulling up roof boards as they are cut or waiting until all are cut to pull them up is a matter of personal preference and has no effect on safety or efficiency.

16. When venting a roof

    a. Vent anywhere along the ridge.

    b. Vent at the hot spot.

    c. Vent as close to the hot spot as can be safely done.

    d. Vent using built-in features (scuttles, skylights, etc.) rather than by cutting a hole in the roof regardless of the location of the hot spot.

17. Smoke ejectors are generally used during

    a. Natural ventilation

    b. Negative-pressure ventilation

    c. Positive-pressure ventilation

    d. All of the above

18. When using positive-pressure ventilation, the pressure inside the building should be

    a. Greater than outside

    b. Less than outside

    c. Equal to the outside pressure

    d. Could be any of the above depending on the direction you want the smoke to travel

19. The positive-pressure blower is placed

    a. Outside the building

    b. Near or in the exit opening

    c. Just inside a suitable entrance

    d. In the position that results in the most effective smoke movement

20. A _____ stream is normally used for hydraulic ventilation.

    a. Fog

    b. Straight

    c. Solid

    d. Combination

Ventilation Operations

Chapter 5

Learning Objectives

- Analyze the utilization of ventilation techniques in several types of free-burning fire situations.

- Realize that ventilation tasks must be performed as part of an overall, coordinated fire attack operation.

- Recognize the limitations of elevators and their use or nonuse during a fire or potential fire incident.

- Understand the term "smoldering fire" and the conditions that may indicate the presence of this situation.

- Be familiar with the term backdraft and its potential impact on the fire ground.

Fire fighters assigned to ventilation operations should be well trained and follow recommended safety guidelines. They must be in full personal protective clothing and self-contained breathing apparatus when performing these tasks. Fire fighters conducting ventilation operations often work above ground level from ground ladders or aerial devices. Members should adhere to all safety regulations while working from this equipment. Ventilation operations are often carried out on roofs or other elevated structures. Fire fighters must pay particular attention to working safely from a flat or pitched roof. It would be helpful if they knew the fundamentals of building construction features. A valuable asset would include a preincident plan and company inspections of the building. Fire fighters must have the proper hand and/or power tools and have been taught how to operate them properly and safely.

Fire fighters must have a working knowledge of vertical- and horizontal-ventilation operations in addition to the concept of positive-pressure and negative-pressure ventilation. They must be able to recognize when ventilation is needed or when it may be an unnecessary risk. Conditions may deteriorate rapidly, and they must be cognizant of changing conditions in their working environment. Proper ventilation saves lives and enhances fire fighter safety.

In the previous chapter, the benefits of proper ventilation and the ways in which fire buildings can be opened to remove accumulated heat and combustion products were discussed. It is not necessary to use every venting technique in every building. The way in which a specific fire building is vented depends on its size, construction, and occupancy; the size and location of the fire; and whether the fire is free burning or smoldering.

This chapter discusses the utilization of venting techniques in several types of fire situations. Not every situation is included, but those that are discussed illustrate how some basic principles may be applied to any structural fire. In a particular situation, ladder company officers and crew members must act on the basis of their knowledge of venting techniques and their sizeup of the situation.

On the fire ground, ventilation duties must be performed as a part of the overall coordinated fire attack operation. Fire fighters working as a team must be able to enter the building to perform the primary search and extinguish the fire. If ventilation is not performed or if improper ventilation tactics are employed, disastrous results could occur. It is imperative that ladder company and engine company members work together in an organized manner and not go off in their own direction or act on their individual perspective. When the fire is free burning, ventilation should begin at the same time as the initial attack or as soon after it as possible. When the fire is smoldering or is suspected to be smoldering, the building must be vented before it is entered. A large segment of this chapter deals with ventilation

Key Points

In a single-story dwelling, ladder crews should open or remove windows close to the fire from either the inside or outside of the building, while the engine crews begin the fire attack.

operations for free-burning fires; the last section discusses how venting techniques are applied to smoldering fires.

One- and Two-Story Dwellings

In most situations, one- and two-story dwellings can be adequately ventilated horizontally through windows. There is usually no need to vent through either natural or forced roof openings unless the fire is in the attic.

One-Story Dwellings

In a single-story dwelling, ladder crews should open or remove windows close to the fire from either the inside or outside of the building while the engine crews begin the fire attack. The first windows to be opened are those through which fire or smoke is pushing out of the structure or through which fire can be seen or heard **Figure 5-1**. This will immediately improve conditions in the dwelling and allow engine crews to advance their hose lines as necessary.

Ladder crews should enter the dwelling and begin the primary search for victims while checking for fire extension. At that time, they should open any other windows needed to complete the ventilation process. The attic or cockloft should be checked for fire spread, especially in the area directly above the fire.

Two-Story Dwellings

If the fire is on the first floor, the first-floor windows closest to the fire should be opened immediately. As noted previously, the first windows to be opened are those that show fire or smoke. The second floor should also be vented above the fire and a search for victims and fire extension begun as soon as ladder crews can enter the dwelling.

If the fire is on the second floor, that floor must be vented first. If conditions warrant, ventilation can be started by ladder crews using the appropriately sized extension ladders placed outside the building. Where possible, the fire floor should also be vented from the inside.

Key Points

If the fire is on the second floor, that floor must be vented first.

Figure 5-1 When fire is seen behind windows or smoke is pushing out around them and fire can be seen or heard, ladder company personnel should open these windows from the outside to assist in gaining entrance.

Key Points

The attic or cockloft should be entered, checked for fire, and vented if necessary. This is especially important where units of a two-family dwelling are side by side, as such dwellings often contain a single common attic across which fire can spread from one unit to the other. Because fire could possibly spread through the dividing wall between side-by-side units, this too should be carefully checked along its full length and height.

Attic Fires

In many dwellings, the attic may be an undivided and open area, completely or partially floored, and often loaded with stored household goods. Attics may be access from exterior stairs in older homes but typically from interior stairs for homes built since the 1970s. Other dwellings have much smaller attics that are lower and only partially usable for storage. This type of attic may be entered by stairs or by climbing a set of pull-down stairs or through a scuttle at the ceiling level inside the house in a hallway or near a stairway.

A working fire in either type of attic should be attacked from within the building, rather than through the windows. To aid engine companies, ladder crews must ventilate the attic. If there is a window at each end of the attic, both windows should be opened or removed from the outside. If an attic has no windows,

it may have built-in louvers at each end for normal ventilation. These are usually located under the peak of the roof and can be removed to accelerate venting. If necessary for adequate venting, the roof should be opened at or near the hot spot above the fire.

Basement Fires

A working fire in the basement of a small dwelling should be ventilated through all available basement openings. In addition, the first floor should be thoroughly vented.

The venting of the basement should be coordinated with the movement of attack lines. If possible, the attack lines should be advanced through both the outside basement entrance and the first-floor basement entrance. If this is done, it must be a coordinated effort to avoid opposing attack lines working against each other. Engine companies should attack the fire from the

Key Points

If there is a window at each end of the attic, both windows should be opened or removed from the outside. If an attic has no windows, it may have built-in louvers at each end for normal ventilation. These are usually located under the peak of the roof and can be removed to accelerate venting. If necessary for adequate venting, the roof should be opened at or near the hot spot above the fire.

Key Points

A working fire in the basement of a small dwelling should be ventilated through all available basement openings. In addition, the first floor should be thoroughly vented.

unburned side of the building whenever possible. Then any other doors and the basement windows can be used for venting. If the fire is attacked through only a single stairway, then any other available openings can be used to vent the basement. By also venting the first floor, ladder crews will aid engine company personnel in positioning their hose lines and advancing them to the basement. If ventilation is not performed quickly and engine companies are unable to gain entry to extinguish the fire, it could travel upward through vertical openings throughout the building and into the attic.

Multiple-Use Residential and Business Buildings

These buildings involve a wide range of sizes, shapes, layouts, and construction materials. Some have stores on the ground floor and apartments or offices above. Older buildings with such combinations of tenancy usually have from two to six stories and are attached to other buildings. Some newer construction shopping centers have one or two stories of offices above the ground-floor retail occupancies.

Older buildings are usually of brick and wood joist construction. Their interior stairways are usually open, and elevator shafts can be open or closed. They probably contain a number of vertical shafts for dumbwaiters, building utilities, and ventilating systems.

With the exception of fire-resistant buildings, these structures and buildings of similar construction, including schools, hospitals, and other institutions, all require the same general ventilation procedures. When a working fire has made considerable headway in such a building, vertical ventilation should be considered at the roof.

Roof Operations

Access   In order to vent the roof, ladder crews must get to it. The quickest way to access the roof is by using an aerial device. If the aerial device must be used for rescue, top-floor venting, and any other priority operations, the ladder crew may need to find an alternative way to the roof. Ladder personnel should use caution if using the interior stairs in the fire building for access to the roof. They could find themselves in an untenable and dangerous position if the fire extends to a corridor or to the stairway itself.

If the fire building abuts a building of the same height, ladder crews can climb interior stairs in that structure to the

Key Points

In order to vent the roof, ladder crews must get to it. The quickest way to access the roof is by using an aerial device.

uninvolved roof and then cross over to the roof of the fire building. If the building is provided with a fire escape, fire fighters may use it if it is not crowded with occupants evacuating the fire **Figure 5-2**. They must, however, make sure that the fire escape they use is safe to do so and has a ladder to the roof.

If there is no other option to gain access to the roof, ladder crews must use ground ladders, aerial ladders, or aerial platforms. An aerial ladder should be placed so that several rungs extend beyond the roof and lighting should be provided if available. This placement, as shown in **Figure 5-3** on page 82, will allow fire fighters operating in smoke and/or darkness to find the ladder when they need it. An aerial platform should be placed so that at least half the width of the basket extends above the roof.

An aerial ladder or platform used for roof access should remain in place until fire fighters leave the roof. If it is needed for rescue or another operation, the new task should be coordinated between command and the affected companies so everyone involved knows that conditions have changed. The aerial device must be returned to its original roof position as soon as that task is completed.

If visibility is poor, ladder crews should probe for the roof with tools long enough to sound it before stepping onto it, as shown in **Figure 5-4a** on page 82. The roof may often be well below the top of its surrounding parapet wall, especially at the front of a building. A fire fighter dropping or jumping down to the roof when it cannot be seen may be seriously injured or killed. When fire fighters reach the roof, by whichever means, they should immediately look for another way off, to be used in an emergency.

Personnel   At least two fire fighters should be sent to the roof for the venting operation. They can work together and keep track of each other. If one is injured, the other will be able to call for help and render assistance. Fire fighters must always work in teams for their own personal safety. If ladder company members are split up, they should be kept with another team, such as an engine company, and work under direct supervision of the officer in charge of that company or sector. Fire fighters within

Key Points

If there is no other option to gain access to the roof, ladder crews must use ground ladders, aerial ladders, or aerial platforms.

Do not use fire building stairways.

Fire escape

Stairs of adjoining building

Figure 5-2 To reach the roof of the fire building when the aerial device is needed for rescue operations, ladder crews can use the fire escape if it has a ladder to the roof or cross over from the roof of an adjoining building of similar height.

a company should have a least one portable radio. If available, every fire fighter should have their own individual radio for safety and communications.

Venting Roof venting should begin with whatever natural openings are available. A roof feature that shows smoke should be opened first, unless fire is directly under the roof. In that case, open over the fire first. Skylights, scuttles, and penthouses should then be quickly opened. The tops of vertical shafts should be checked; a shaft should be opened if heat or smoke can be detected around it, as shown in **Figure 5-5** on page 83.

Once the roof has been vented, the top floor should be opened. Ladder crews could use hand tools to break the glass of the windows from the roof, if this task can be accomplished safely. Fire escapes and front or rear porches or balconies are other positions where ventilation could be performed. If conditions permit, ladder crews could work their way down to the top floor from the roof using an interior stairway or a scuttle that has a ladder. They then could open windows from the inside and, at the same time, begin a primary search of the top floor.

Key Points

At least two fire fighters should be sent to the roof for the venting operation. They can work together and keep track of each other. If one is injured, the other will be able to call for help and render assistance.

Key Points

Roof venting should begin with whatever natural openings are available. A roof feature that shows smoke should be opened first, unless fire is directly under the roof. In that case, open over the fire first.

Figure 5-3 An aerial ladder should be placed several rungs above the roof edge so it can be located easily in smoke or darkness.

Key Points

After the roof has been vented, the top floor should be opened. Ladder crews could use hand tools to break the glass of the windows from the roof, if this task can be accomplished safely.

Key Points

Ladder personnel should not attempt to enter the building from the roof if the top floor is heavily charged with smoke. Instead, they should leave the roof and open the top-floor windows from aerial ladders or platforms.

Ladder personnel should not attempt to enter the building from the roof if the top floor is heavily charged with smoke. Instead, they should leave the roof and open the top-floor windows from aerial ladders or platforms. Fire fighters using ground ladders could ventilate the top-floor windows of shorter buildings. The top floor should be searched as soon as it is tenable.

The floor just above the fire floor must be thoroughly vented. There are two important reasons for this: A primary search for victims must be completed, and the floor must be checked for

Figure 5-4 Before getting off the aerial device, fire fighters must determine the actual location and condition of the roof to avoid serious injury.

vertical fire extension, which needs to be controlled so that the fire does not travel vertically to higher floors. The fire floor must be vented to allow for the advancement of hose lines and a search for victims, as well as for fire fighter safety **Figure 5-6** (page 84).

Ground-Floor Stores

The stores on the ground floor of a combination commercial/residential occupancy are often structurally separated from the upper floors; that is, there may be no stairways or shafts extending through the rest of the building from the stores. In spite of this separation, smoke and other combustion products will work their way through the structure, especially if the store is heavily involved with fire.

It may be necessary to ventilate the roof, the top floor, and the floor above the fire, even when the fire is in a ground-floor store. The store itself should be thoroughly vented and the fire attacked quickly and aggressively. Utility shafts that serve the store and the upper floors, or other ground-floor stores, must be opened to deter fire spread.

Adjoining Buildings

When a multiple occupancy is well involved and is one in a row of similar structures, the adjoining occupancies on either side should be vented and the area of the cocklofts or attics checked for extending fire. This is especially important if the top floors of the fire building are involved.

Key Points

It may be necessary to ventilate the roof, the top floor, and the floor above the fire, even when the fire is in a ground-floor store.

Key Points

When a multiple occupancy is well involved and is one in of a row of similar structures, the adjoining occupancied buildings on either side should be vented and the area of the their cocklofts or attics checked for extending fire.

Figure 5-5 Once on the roof, ladder crews should use natural openings to begin venting, unless the fire is directly under the roof. In that case, open over the fire first.

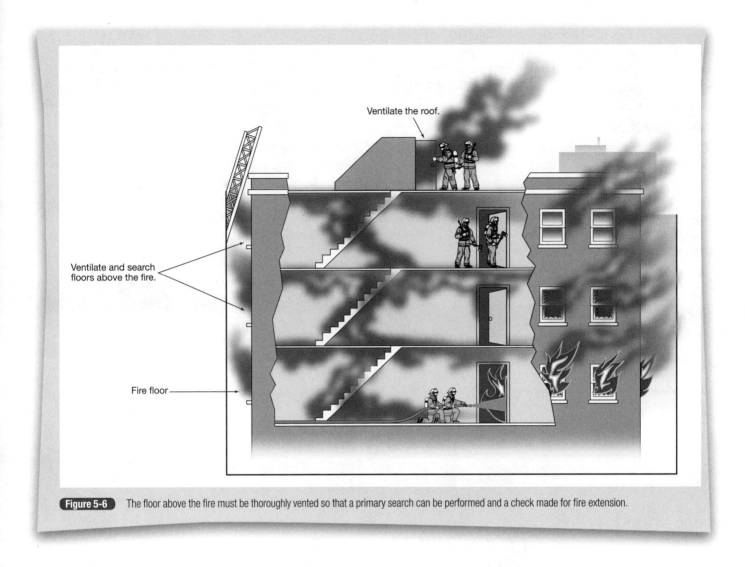

Ventilate the roof.

Ventilate and search floors above the fire.

Fire floor

Figure 5-6   The floor above the fire must be thoroughly vented so that a primary search can be performed and a check made for fire extension.

The adjoining building should be vented through skylights, scuttles, and penthouses, if any of these features are part of the structure, to keep the damage to a minimum; however, if the cockloft is not vented when these features are opened, the boxed-in areas below them will have to be forced open. If the fire has extended into an adjoining occupancy, this venting will release smoke and gases and slow the fire's spread there.

The exposure fire then should be completely vented, with a roof opening at the hot spot if necessary. The extension of fire into another occupancy must be reported to command immediately so that additional resources may be deployed to contain the exposure fire.

Key Points

The extension of fire into another occupancy must be reported to command immediately so that additional resources may be deployed to contain the exposure fire.

Shopping Centers, Row Stores, and Other One-Story Buildings

Fires involving shopping centers, old row stores, factories, warehouses, and other large one-story structures sometimes cause excessive damage simply because they have not been ventilated properly. It is true that these buildings present only a limited problem in terms of vertical fire spread, but they are extremely vulnerable to horizontal fire spread.

Factories and warehouses usually contain large open areas through which fire can spread quickly. Shopping centers and row stores may often have an unbroken cockloft and basement area, which may possibly serve many stores. In addition, there may be hanging ceilings, and unsealed openings in fire walls for utility pipes, electrical, and telephone services. All of these features provide channels for horizontal fire spread from store to store.

Proper ventilation can stop or slow the spread of fire in one-story structures. Preincident planning and inspection will assist ladder companies in determining how the buildings in their district should be vented.

Roof Operations

Ladder companies arriving at a working fire in a large one-story structure should always assume that there are no fire walls between stores or building sections. Ladder crews should be sent to the roof in preparation for ventilation. There is usually no problem in reaching the roof. An appropriate ground ladder can be placed for access to the roof of an adjoining occupancy or to the roof of the involved building if it is wide enough. Ladders should be kept away from windows and/or door openings. A ladder should not be located here if fire vents from these areas.

It also keeps the ladders out of the way of engine companies advancing their attack lines.

A **natural roof feature** should be used for the first roof opening only if it is close enough to the hot spot; otherwise, the roof should be cut open at the hot spot from the fire or as close to it as safety permits. After one opening has been made over the main body of fire, natural openings can be used to complete the ventilation job **Figure 5-7**.

Modern one-story warehouses are, in many cases, practically windowless. Roof venting operations must be effective to allow for proper fire attack. These buildings often have several skylights, usually of the bubble type, installed across the roof. Those directly over the fire should be opened first, if possible. Then, as shown in **Figure 5-8**, other skylights in the area should be opened. As in any other structure, if there are no natural openings above the fire, the roof should be opened at the hot spot or as close to it as safety permits.

Fire fighters must have a good understanding of roof construction and the effect of fire on the various roof types. The crews should be aware of any roofs in their area that might

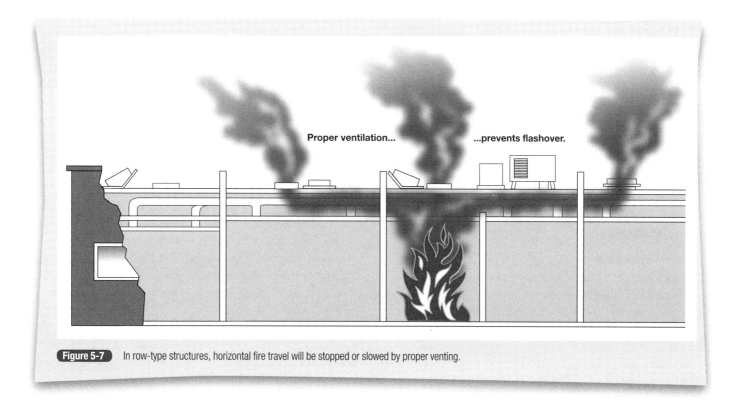

Proper ventilation...   ...prevents flashover.

Figure 5-7   In row-type structures, horizontal fire travel will be stopped or slowed by proper venting.

Figure 5-8   In a one-story warehouse, the roof directly over a fire should be vented first then fire fighters should work outward from that area to vent other parts of the roof.

Key Points

Modern one-story warehouses are, in many cases, practically windowless. Roof venting operations must be effective to allow for proper fire attack.

Key Points

If the roof feels spongy or is sagging, the steel could be warped and, therefore, weakened. Fire fighters should always avoid such areas.

be particularly hazardous. One word of caution: The steel roof girders and joists in many warehouses, garages, and large one-story superstores are exposed from below, and thus, there is no protective layer between the steel and the heat of the fire. Fire fighters reaching such a roof should check its condition carefully. If the roof feels spongy or is sagging, the steel could be warped and, therefore, weakened. Fire fighters should never enter such areas. Although they should open the roof as close to the fire as safety permits, they should not endanger themselves to do so **Figure 5-9** .

Attached Occupancies

After the roof of the fire building is properly opened, the roofs of attached buildings should be opened through natural roof features as needed. This will allow ladder crews to determine whether the fire has spread into the cocklofts of the attached occupancies, which usually are other stores. This also will allow accumulated gases to escape so that they will not ignite when the ceilings of exposures are pulled down to check for fire extension.

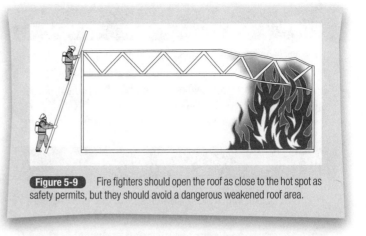

Figure 5-9   Fire fighters should open the roof as close to the hot spot as safety permits, but they should avoid a dangerous weakened roof area.

Key Points

After the roof of the fire building is properly opened, the roofs of attached buildings should be opened through natural roof features as needed.

If fire is found in an attached exposure, its roof should be opened over the fire. Then its ceiling should be pulled down, and the fire attacked from below.

Ground-Level Ventilation

When necessary, a store can be ventilated at ground level through its front display windows and through rear windows and doors. Some older display windows are constructed with a large main window below and window lights above. The window lights, being made of thinner glass than the main window, are easier to remove and cheaper to replace. They should be removed first. Because they are located near the ceiling and near the greatest concentration of heat and smoke, they might provide sufficient venting. If not, then the main window should be removed. There should be no hesitation in removing the display window if fire conditions dictate.

Basement Fires in Large Structures

A basement is the worst possible place for a working fire because it exposes the entire building and all of its occupants. Because heat, smoke, and gases will travel vertically throughout the building, spreading fire and overcoming occupants, it is imperative that the building be ventilated quickly and properly. The larger the involved area, the greater the ventilation effort is required.

As a start, the basement, the first floor, and any shafts that lead up through the roof should be vented. The release of accumulated heat, smoke, and gases from the basement and first floor will reduce the chance of vertical extension.

Key Points

Any opening into the basement, such as a door, window, chute cover, bulkhead, or sidewalk door, can be used for venting. If possible, ventilation openings should be opposite those being used for the fire attack.

Basement Venting

Any opening into the basement, such as a door, window, chute cover, bulkhead, or sidewalk door, can be used for venting. If possible, ventilation openings should be opposite those being used for fire attack If the fire is being attacked from a rear basement entrance, then the basement should be vented from the front and/or the sides if the building is detached. If the fire must be attacked from the front of the building, ladder crews should ventilate at the rear. If a single available basement entrance must be used for the fire attack, the area should be vented through available basement windows. Keep in mind that whenever possible the fire should be attacked from the unburned side.

A basement with only one entrance, which is being used for the fire attack, and no windows presents a serious venting problem. If the fire is extinguished quickly, venting the first floor may suffice to clear the basement. If not, the first floor should be cut through to open it to the basement. Windows should then be opened or removed to draw the products of combustion up from the basement **Figure 5-10**. Smoke ejectors

Figure 5-10 If the attack does not have an immediate effect on the fire, the first floor should be opened just inside the windows and the windows opened or removed.

Key Points

A basement with only one entrance, which is being used for the fire attack, and no windows presents a serious venting problem. If the fire is extinguished quickly, venting the first floor may suffice to clear the basement.

Key Points

Where basement windows do exist, they might be below grade and covered with steel grates. If the grate openings are wide enough to admit the end of a tool, the window glass can be broken out through the grate. If not, the grate must be removed before the window is broken.

or positive-pressure fans could be used to assist in the ventilation process.

Where basement windows do exist, they might be below grade and covered with steel grates. If the grate openings are wide enough to admit the end of a tool, the window glass can be broken through the grate. If not, the grate must be removed before the window is broken. For this, place a flathead axe or the adz end of a Halligan or Kelly tool between the grate and the concrete around it at the narrow end or at a corner. Drive in the tool and use it to pry up the grate.

Storefront Walls When a basement fire is being attacked from the rear of the building or from an inside stairway, it is important that the basement be ventilated. If there are no basement windows or doors at the front of the fire building but the ground floor is occupied by a store, open the low wall below the display window to vent the basement. Often, the display window wall opens directly to the basement or to the basement ceiling, which can quickly be knocked down into the basement **Figure 5-11**. Opening the wall under display windows can be an effective venting operation. First remove the glass and then the wall below the window. Check for a ceiling below that level.

Before the wall is opened, the display window should be removed with a pike pole, plaster hook, or similar tool. This ensures the display window does not break and shower glass on fire fighters while they are opening the wall. It also will help to vent the first floor.

With the display window removed, the wall can be opened. The procedure to be used depends on the materials involved. It might be necessary to cut through wood sheathing, pry steel panels off wood framing, or knock out brick veneer depending on the construction type of the building.

If the wall below the display window opens directly to the basement, smoke should pour out as soon as the wall is cut through.

Figure 5-11 Fire fighters can open the wall under display windows as an effective venting operation.

If not, there could be a basement ceiling between the basement and the wall. The ceiling can be pushed down with a pike pole or similar tool. The openings in the display window wall and the basement ceiling should be made as large as possible for effective venting.

The display-window wall might be constructed of a material that is very difficult to remove, such as glass block. In such cases, the display-window floor can be opened to ventilate the basement. Again, the display window is first removed. Then the items in the window should be moved out of the way and the display-window floor cut through near the front of the display area. Finally, the basement ceiling should be opened, if necessary. Smoke ejectors or positive-pressure fans can be used to assist in the ventilation process as smoke emanates from the ventilation opening; however, the smoke ejectors should not be placed directly over the hole until the fire is knocked down **Figure 5-12**.

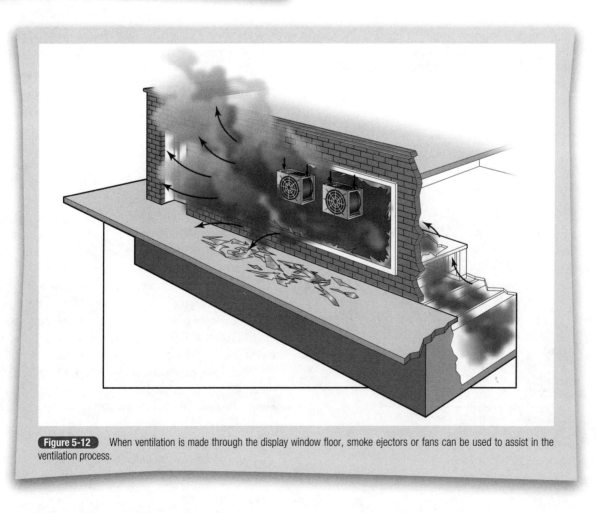

Figure 5-12 When ventilation is made through the display window floor, smoke ejectors or fans can be used to assist in the ventilation process.

Key Points

The display-window wall might be constructed of a material that is very difficult to remove, such as glass block. In such cases, the display-window floor can be opened to ventilate the basement.

Key Points

When the fire is near the front of a long basement and the store has two display windows, both display-window walls and floors should be opened.

When the fire is near the front of a long basement and the store has two display windows, both display-window walls and floors should be opened. One opening can be used for fire attack with hose lines and the other for venting. It is usually necessary to vent the rear of the basement at the same time.

Other Openings   Any opening that will help ventilate the basement should be used to permit the advancement of attack lines and lessen the chance of vertical and lateral fire spread.

First-Floor Venting

The first floor of the fire building should be vented through windows to remove combustion products that have seeped up from the basement. This will aid engine crews who might be advancing attack lines to the basement, through interior stairways, or to the first floor. The venting will also increase the effectiveness of search operations and reduce the chance of fire spread to the first floor.

Roof Venting

Shafts running through the building must be opened at the roof to release combustion products convected up to that level. This will help vent the basement and will prevent smoke accumulation, ignition of the upper part of the building, and fire spread to the tops of attached buildings. If vertical shafts are not opened at a

Key Points

The first floor of the fire building should be vented through windows to remove combustion products that have seeped up from the basement.

Key Points

Shafts running through the building must be opened at the roof to release combustion products convected up to that level.

serious basement incident, fire fighters probably will be faced with a second fire at the top of the building.

Fire-Resistant Structures

Fire-resistant construction is used in many structures classified as low, medium, and high rise. The main structural members of these buildings have special coatings to protect them from heat. Other construction features are designed to retain the heat of a fire in as small an area as possible. Doors to apartments and offices are made of steel, and most codes require that they be self-closing. Floors are concrete. Stairways and elevator shafts are enclosed. Utility shafts and heating and air conditioning ducts have high protective ratings. The interior stairways are constructed to serve as enclosed fire escapes.

Although these construction features restrict fire spread, they tend to result in extremely high temperatures in the immediate area of a fire and in very heavy accumulations of smoke and fire gases. A number of deaths and injuries can be traced to the spread of smoke and fire gases throughout these fire-resistant buildings, many by way of the heating, ventilation, and air conditioning systems combined with a lack of proper ventilation.

A fire-resistant structure cannot be vented in the same way as a more standard structure—at least initially, when the building is occupied. For example, the enclosed stairways must not be opened to smoke-filled corridors while they are being used as escape routes; therefore, they cannot be used for venting.

Window Venting

The best way to ventilate an occupied fire-resistant building is through its windows. To do so, ladder crews must force entrance into the rooms, apartments, or offices on both sides of halls and corridors. The entry doors should be propped open and as many windows as possible opened in each unit. If the fire has

Key Points

A number of deaths and injuries can be traced to the spread of smoke and fire gases throughout fire-resistant buildings, many by way of the heating, ventilation, and air conditioning systems combined with a lack of proper ventilation.

Key Points

Fire-resistant construction is used in many structures classified as low, medium, and high rise. The main structural members of these buildings have special coatings to protect them from heat. Other construction features are designed to retain the heat of a fire in as small an area as possible.

Key Points

A fire-resistant structure cannot be vented in the same way as a more standard structure—at least initially, when the building is occupied.

Key Points

The best way to ventilate an occupied fire-resistant building is through its windows.

Key Points

By opening units on both sides of the corridor, ladder crews can create cross-ventilation that will quickly begin to clear the area of combustion products.

control of a corridor, ladder crews should advance along with the engine crews on the attack lines, force entry into a unit as they come to it, open the windows in the unit, and then rejoin the engine crews **Figure 5-13**. This procedure should only be accomplished when the fire will not be drawn down the corridor and into apartments or offices.

By opening units on both sides of the corridor, ladder crews can create cross-ventilation that will quickly begin to clear the area of combustion products. If necessary, smoke ejectors or positive-pressure ventilation fans, if used properly, can be used to aid in the venting. As windows are opened, the fans should be placed to draw or push the smoke outside.

The fire floor should be vented first. As soon thereafter as possible or at the same time, if personnel are available, the floor above the fire floor should be vented and a primary search conducted for victims. Despite the fire-resistant construction, smoke can seep up through the building. Engine companies should advance hose lines to the floor above the fire to keep it from extending vertically from window to window or through other vertical openings.

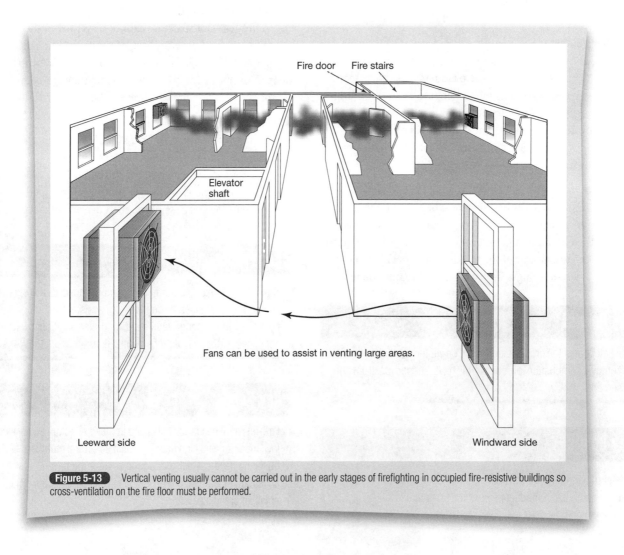

Fire door   Fire stairs

Elevator shaft

Fans can be used to assist in venting large areas.

Leeward side   Windward side

Figure 5-13   Vertical venting usually cannot be carried out in the early stages of firefighting in occupied fire-resistive buildings so cross-ventilation on the fire floor must be performed.

Key Points

The fire floor should be vented first. As soon thereafter as possible or at the same time, if personnel are available, the floor above the fire floor should be vented and a primary search conducted for victims.

Key Points

If the building is unoccupied or has been completely evacuated, doors leading from corridors to stairways can be opened. The stairway shafts will draw smoke from the corridors and provide vertical venting.

Stairway Venting

If the building is unoccupied or has been completely evacuated, doors leading from corridors to stairways can be opened. The stairway shafts will draw smoke from the corridors and provide vertical venting. In many cases, the penthouses above the stairway are fitted with ventilation openings, and ladder crews need not climb to the roof to vent them. Obviously, if doors need to be opened to the outside for ventilation members will have to perform this task. Such details should be determined through preincident planning and inspection.

The larger the fire, the more pent-up heat and smoke will be encountered. Fire in a large open area in a fire-resistant structure will require the use of larger attack lines. Along with the smoke and heat, this will place great physical strain on engine company personnel, and thus, ladder crews must quickly remove as much heat and smoke as possible by venting through all available openings.

Other Ladder Duties

It is extremely important in a fire-resistant structure that the fire floor and the floor above it be vented, searched, and checked for fire spread. Ladder crews must work closely with engine crews by ventilating to keep the hose lines moving.

In most small fires in these structures, rescues will be made from the apartment, office, or work area in which the

Key Points

It is extremely important in a fire-resistant structure that the fire floor and the floor above it be vented, searched, and checked for fire spread.

Key Points

While proceeding with ventilation assignments, ladder crews must search every area through which they pass, including lobbies, elevator alcoves, corridors, the fire floor, and the floors above it.

Key Points

Venting duties are secondary to the rescue of victims who have become lost, trapped, or overcome by smoke and fire gases.

fire originated; however, if the fire, smoke, and products of combustion have spread into the corridors, and upper floors, frightened occupants, who may have done better to stay in their apartments or offices, might attempt to reach an exit but collapse en route. Therefore, while proceeding with ventilation assignments, ladder crews must search every area through which they pass. This includes lobbies, elevator alcoves, corridors, the fire floor, and the floors above it. Venting duties are secondary to the rescue of victims who have become lost, trapped, or overcome by smoke and fire gases.

Using Elevators to Approach the Fire Floor

Fire departments must have a standard operating guideline for the use of elevators during a fire or potential fire incident. They must know what they can and cannot do with an elevator during a fire incident. Fire fighters should be familiar with the different elevators that are manufactured and how they operate. They should know if their jurisdiction requires automatic recall of elevators to a landing zone and if the elevator is equipped for fire service operation.

If the elevators have lobby controls, their locations and proper operation should be known. Fire fighters should know how to obtain and keep control of elevators. Because elevators and controls vary greatly from one manufacturer to another, meetings with various manufacturers and building personnel are recommended.

Fire departments should have strict guidelines concerning the use of elevators. Many departments disallow the use of elevators without the fire service feature and disallow elevators to be used for incidents on lower floors.

Key Points

Fire departments must have a standard operating guideline for the use of elevators during a fire or potential fire incident.

Key Points

Fire fighters should be familiar with the different elevators that are manufactured and how they operate.

In some cases, fire fighters intending to take the elevator to a floor below the fire have been taken directly to the fire floor by mistake.

No fire fighter should ever be allowed to enter an elevator without full PPE, including SCBA, a portable radio, and tools while responding to a fire or potential fire incident, such as a fire alarm sounding.

If an elevator is available, it can be used to transport ladder crews and their equipment to the vicinity of the fire floor in both fire-resistant and standard structures; however, the elevator should never be taken to the fire floor itself. If it is, those in the elevator might be exposed to flames or excessive heat when the door opens. It is much safer to take the elevator to at least two to four floors below the fire floor and walk up.

When the location of the fire is difficult to judge and might have been erroneously reported, fire fighters should get off the elevator several floors below the supposed fire floor, even if smoke or heat detectors are used to determine the fire floor.

Ladder crews can also use the elevator as an equipment carrier after personnel are in position. Tools, fans, lights, SCBA cylinders, and engine-company EMS equipment can be sent up from the ground floors, and an equipment pool established by command on a designated floor below the fire floor.

In an elevator emergency, ladder crews can use forcible-entry tools, which they should be carrying, to force elevator doors open or shut as required, to remove overhead escape panels, and if necessary, to chop their way out of the elevator shaft. If any of these actions need to be taken, it certainly will be a difficult operation for the fire crews.

In structures without automatic recall, it is often impossible for the fire department to obtain elevators immediately upon arrival since tenants will be using them to escape despite warning notices to the contrary. Power outages may also render elevators useless. In this case, ladder crews must use stairways to get up to the fire floor. Fire departments should have high-rise bags or other similar methods of transporting tools and equipment to the upper floors. Strict adherence to department policies and guides must be followed. All safety equipment and full PPE must be used.

Smoldering Fires

On rare occasions, fire companies responding to an alarm will encounter a fire that is not or does not appear to be free burning. There will be plenty of smoke to indicate that there is a fire, but without visible flames. This condition exists because the fire is being deprived of sufficient oxygen to maintain open combustion. In such a situation, fire fighters must assume that they have come on a smoldering fire, which if not handled properly could lead to a serious situation.

Indications

One or more of the following conditions may indicate a smoldering fire:

- A lot of smoke is visible, but no open fire can be seen or heard.
- Smoke is rising rapidly from the building, an indication that it is hot. (However, humid weather or an atmospheric inversion may be holding down the smoke.)
- Smoke is leaving the building in puffs or at intervals.
- Some smoke is being drawn back into the building around windows, doors, and eaves.
- Smoke is a gray or yellow color.
- Pressurized smoke is coming from cracks or openings.
- Windows are stained brown or darkened from the intense heat.
- Window glass is hot to the touch.

Occasionally, one or more window panes may be broken by the heat from the inside of the building. A small rim of fire can appear around the edges of the broken glass, where the oxygen content is high enough to ignite some of the gases. This indicates that a backdraft is imminent. Whenever any of these conditions appears to be present, the fire must be handled as a smoldering fire for the safety of fire fighters and for proper firefighting operations.

A smoldering fire has sufficient heat and fuel to become free burning. The heat comes from the fire, which was probably burning freely at one time. The fuel is mainly carbon monoxide gas from the original fire and from the smoldering fire, which originated in the contents of the building and, perhaps, the building itself. The carbon monoxide has filled the building and surrounded the smoldering fire, thus cutting off its oxygen supply. Lack of oxygen keeps flames from developing, but the fire smolders and produces intense heat. A smoldering fire needs only oxygen to burst into flame, with no special circumstances involved. A fire can be smoldering in a building of any size or type or, in some cases, in only one area of a large structure.

Backdraft

A smoldering fire must be ventilated before it is attacked; that is, carbon monoxide must be removed from the building before air is allowed to enter. The addition of any oxygen to the heat and fuel will lead to immediate ignition.

This sudden ignition can take any of several forms. In one situation, the gases and heated combustibles might simply burst into flames, engulfing the building or a part of it in fire. In another, the force of the ignition might be enough to blow windows, doors, and fire fighters out of the building **Figure 5-14**. There could also be an explosion strong enough to cause structural damage to the building and injury to fire fighters. Just what will happen cannot be determined beforehand, but it is certain that the addition of oxygen will cause some sort of ignition. That ignition is referred to as **backdraft.**

Venting

While the fire is smoldering, the carbon monoxide, other heated gases, hot air, and smoke will have been convected upward and will be collecting at the top of the structure. An opening must be made as high on the building as is safely possible to release these gases and allow them to move out of the structure (Figure 5-14). This is the same principle used to vent free-burning fires. The difference is that with smoldering fires the venting must be done *before* the building is entered to relieve the explosive situation and reduce the chance of backdraft. If this sequence is not followed, air entering the building with fire fighters will cause immediate ignition.

It is important to ventilate fully and in the right places to ensure that the hot gases are dispersed. On one hand, ventilation should not be rushed or haphazard. On the other, it should be completed promptly. Heat, smoke, and gases have been building up inside the structure, and thus, it is possible for these

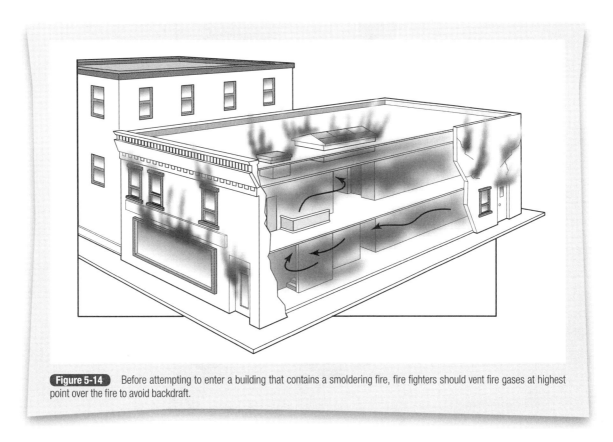

Figure 5-14 Before attempting to enter a building that contains a smoldering fire, fire fighters should vent fire gases at highest point over the fire to avoid backdraft.

combustion products themselves to break that window glass and allow air to enter before ventilation is completed. Again, the result would be sudden ignition; this time, however, fire fighters would be near the building.

Natural roof openings can be used to vent the building, and holes can be cut if necessary. Ladder crews going to the roof will usually be able to determine which natural features to open first by noting the amount of smoke pushing out around them. If roof venting seems particularly dangerous, break the tops of the highest windows using extreme caution for personnel safety. A solid-stream hose line may be able to perform the task depending on the type of window. After the top of the building is opened, the remainder can be vented as necessary.

Because a backdraft can take the form of a violent explosion, releasing a tremendous blast of fire and heat, fire fighters must avoid approaching the building directly. Their approach should be made either from an oblique angle or parallel to the building,

as shown in **Figure 5-15**. This is especially important when working at fires in stores or other structures with large glass areas, since these are the weakest points in the building. Fire fighters should not get directly in front of the building until venting has been accomplished. Many fire fighters caught in such a blast have suffered severe burns and critical injuries from flying glass and debris.

This operation requires a coordinated fire attack. Attack lines should be charged and ready for use during the ventilation of the building. Crews on the hose lines should be in safe positions, protected from flying glass, and ready to enter the building as soon as venting is completed. Likewise, if there is any possibility of a backdraft, fire apparatus should not be positioned in a direct line with the building, especially if there are large glass areas at the street level. After the fire is ventilated, it will burn freely. It can then be attacked in the same way as any other free-burning fire, although its size might not immediately be known.

Key Points

While the fire is smoldering, the carbon monoxide, other heated gases, hot air, and smoke will have been convected upward and will be collecting at the top of the structure. An opening must be made as high on the building as is safely possible to release these gases and allow them to move out of the structure.

Key Points

Because a backdraft can take the form of a violent explosion, releasing a tremendous blast of fire and heat, fire fighters must avoid approaching the building directly. Their approach should be made either from an oblique angle or parallel to the building.

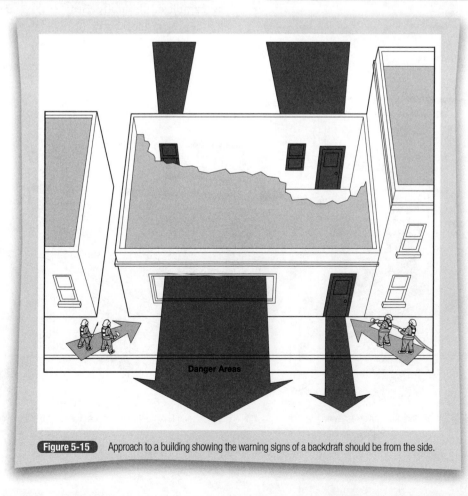

Danger Areas

Figure 5-15 Approach to a building showing the warning signs of a backdraft should be from the side.

Chief Concepts

- Every structural fire must be ventilated. A free-burning fire should be ventilated in coordination with other firefighting operations. A smoldering fire must be ventilated as high as possible on the fire building before the building is entered.
- In one- and two-story dwellings, ventilation operations can consist solely of performing horizontal ventilation, opening windows on the fire floor and on the floor above the fire, if there is one. In larger structures, ladder crews might have to open the roof and the top floor, as well as the fire floor and the floor above it, in order to ventilate properly. In any building, the attic or cockloft must be checked for fire extension and vented if necessary. The tops of attached exposures should also be checked and vented. Basement fires require quick and thorough venting of the basement, the ground floor, and any shafts that extend through the full height of the building.
- Fire-resistant buildings are vented mainly through windows on the fire floor and the floor above it. Because these buildings are designed to retain heat and smoke, ladder crews must perform a primary search for overcome victims as they perform their venting duties.

Key Terms

Backdraft: The sudden explosive ignition of fire gases when oxygen is introduced into a superheated space previously deprived of oxygen.

Natural roof features: Skylights, scuttles, and penthouses and vertical shafts are examples of natural roof features that can be used to ventilate the building.

1. Ventilation should begin at the same time as the initial attack or as soon after as possible when the fire is
   a. Smoldering
   b. Free burning
   c. In the incipient stage
   d. Showing from the roof

2. Ventilation should begin before the building is entered when the fire is
   a. Smoldering
   b. Free burning
   c. In the incipient stage
   d. Showing from the roof

3. One- or two-story dwellings can be adequately vented
   a. Horizontally through windows
   b. Vertically by cutting an opening in the roof
   c. Vertically by removing roof openings
   d. Horizontally through windows unless the fire is in the attic

4. To ventilate an attic fire in a two-story dwelling
   a. Open the roof near the hot spot.
   b. Remove or open windows on each end of the attic.
   c. Remove built-in louvers at each end of the attic.
   d. Any of the above could be correct depending on the attic.

5. When using an aerial ladder to gain access to the roof
   a. The ladder should be placed so the top rung is even with the top of the roof.
   b. One full rung should extend beyond the roof line.
   c. Several rungs should extend beyond the roof.
   d. The number of rungs to be placed above the roof varies depending on the height of the building and parapet wall.

6. When venting the roof
   a. All ladder company members should work together.
   b. At least three fire fighters should be assigned to the roof with at least one member at the location of roof access.
   c. At least two fire fighters should be sent to the roof.
   d. A single fire fighter can safely work on the roof for a short period of time as long as visibility is good.

7. A fire has extended from the building of origin to the cockloft of an adjoining store. The proper fire attack would involve
   a. Opening the roof of the exposure and operate a fog pattern into the roof opening of the exposure only
   b. Opening the roof of the exposure, pulling the ceiling and attacking the fire from below

   c. Opening the roof of the fire building and attacking the fire by placing ladders into the cockloft and attacking from the unburned side of the cockloft
   d. Do not open the roof of the fire building or exposure but pull the ceiling and attack the fire from below.

8. There is a working fire in the basement of a large structure with only one opening to the basement that is being used to attack the fire. The best venting option would be to
   a. Vent through the one opening to the first floor and then attack the fire after the heavy smoke has been relieved.
   b. Open the floor between the basement and the first floor directly over the fire if possible and then open all windows on the first floor.
   c. Open the floor between the basement and the first floor near a first floor window and vent out the first floor window.
   d. Do no vent this fire and attack via the one available opening.

9. Fire-resistive construction
   a. Restricts fire spread
   b. Results in extremely high temperatures in the fire area
   c. Results in heavy accumulations of smoke and fire gases
   d. All of the above

10. The best way to ventilate an occupied fire-resistive structure is
    a. At the top of the stairways, particularly when there is a scuttle opening to the roof
    b. Using vertical shafts that serve the entire building
    c. By any vertical means
    d. Through the windows

11. When venting an occupied fire-resistive building, start
    a. By venting the fire floor
    b. By venting the floor above the fire
    c. By venting the hallway and stairway
    d. At the top and work down

12. Elevator controls are
    a. Widely varied from one manufacturer to another
    b. Standardized with a few minor differences
    c. Standardized in fire service mode, but differ in other ways
    d. Standardized for passenger elevators, but vary widely for freight elevators

13. A lack of _____ keeps a smoldering fire from bursting into a free-burning fire.

   a. Oxygen

   b. Fuel

   c. Temperature

   d. Heat

14. Which of the following are signs of a smoldering fire? Circle all that apply.

   a. Smoke is a gray or yellow color.

   b. Cool smoke that does not rise.

   c. Smoke is puffing from building.

   d. Smoke is being drawn back into the building.

   e. Smoke is pressurized.

   f. Open flame is visible.

   g. Exit gases are mainly carbon dioxide and water vapor producing a white smoke.

   h. Windows are stained brown or darkened from intense heat.

15. When indications of a smoldering fire are visible from the exterior

   a. Vent from the highest point possible before entering the building.

   b. Vent from the highest point possible once a charged attack line is in position on the interior.

   c. Vent from the lowest possible point before entering the building.

   d. Vent from the lowest possible point once a charged attack line is in position and charged on the interior.

Checking for Fire Extension

Chapter 6

Learning Objectives

- Identify the two categories of fire exposure.

- Recognize the importance of protecting building components or areas from fire extension.

- Assess the importance of checking interior exposures for vertical and horizontal fire spread.

- Assess the importance of checking exterior exposures to prevent the ignition of the exposure or slow the spread of fire.

Fire fighters must pay attention to their safety when checking for fire extension within a fire building or an exposure. They must be aware of their surrounding at all times. Fire can spread quickly through areas of a building not originally involved in fire. Walls, ceilings, attic areas, and any other vertical or horizontal space, concealed or open, are avenues for fire spread in the form of direct flame impingement, embers, radiant heat, or the transfer of heat by convection or conduction.

Dependent on the severity of the fire situation, additional fire fighters may need to be assigned to this operation. The faster the fire is contained and areas are checked for fire extension to prevent additional damage, the job of the fire fighter will become less hazardous. Fire extension could hamper or curtail search-and-rescue operations and cut off the primary means of egress for both fire fighters and occupants.

Fire fighters assigned to this task should be in full personal protective clothing, including SCBA, have radios for communications, possess the proper tools and equipment, and preferably be protected by a hose line when performing their duties. This is an occasion when a preincident plan and the company's familiarization of the building become invaluable. The plan should include details regarding the building's construction features, layout, contents, special hazards, and fire protection systems. With this information and observations made during the incident, the fire's route of travel may be anticipated. Fire fighters can be assigned to areas where the fire is likely to extend and prevent the spread of fire within the building or to other exposed buildings.

As a basic objective of firefighting operations, exposure protection is second only to rescue in terms of importance. Exposures are structures or parts of structures not involved in fire but in danger of becoming involved and must be protected to minimize the danger to their occupants and to contain the fire. The major contribution of ladder companies to exposure protection activities is to check carefully and thoroughly for fire spread.

Exposures are generally classified into two categories according to their location relative to the fire structure: external (exterior) or internal (interior). External or exterior exposures are those that could become involved if the fire spread from another building across some open area. Structures near, but detached from, the fire structure are exterior exposures **Figure 6-1**. Wings of the fire structure that are in danger of fire spread across an open courtyard or an air shaft may be considered exterior exposures.

Figure 6-1 External or exterior exposures are structures that are near, but detached from, the fire building and could become involved if the fire spreads.

Internal or interior exposures are those to which fire can spread from within the fire structure. They include floors in the fire building above and below the fire floor as well as structures attached to the fire building.

The danger to exterior exposures is usually obvious. Direct flame impingement, radiant heat, or burning embers being carried to an exposed structure by convection currents and winds. Radiant heat itself cannot be seen, but it can be felt. Its effects on exposures are obvious—blistering paint, surface discoloration, cracking glass, visible vaporization (gas production), and ignition at various points of a building are indications of the action of radiated heat.

Interior exposures are not so obvious. Ladder crews must seek out hidden channels through which fire can spread. At times less spectacular than exterior exposure fires, interior exposure fires can be much harder to extinguish.

As in every phase of firefighting, preincident planning and area and building inspections are important components of exposure protection. Inspections will help locate exposure hazards, including conditions or situations in or around a building that could promote the spread of fire. Fire fighters should be familiar with their response area and be observant of neighborhoods in which the spread of fire is especially likely. Preincident planning and inspections will help ensure that procedures are in place to dispatch sufficient apparatus and personnel on the first alarm to protect interior and exterior exposures.

When a fire obviously poses an extreme exposure hazard, first-alarm ladder company response should be increased over normal assignments. The additional fire fighters will be necessary so that exposure protection operations can be carried out in the shortest time. In addition to other fireground duties, ladder crews will be required to investigate and determine whether there are any secondary fires extending through attached or detached structures. They may need to force entry into exterior and interior exposures so that engine companies can attack spreading fire and to open up interior channels through which the fire may be

> ## Key Points
>
> Fire fighters should be familiar with their district and be observant of areas or neighborhoods in which the spread of fire is especially likely.

> ## Key Points
>
> When a fire obviously poses an extreme exposure hazard, first-alarm ladder company response should be increased over normal assignments.

> ## Key Points
>
> Ladder companies must check interior exposures to determine where the fire is located and to keep it from spreading to uninvolved areas.

spreading. This last task is perhaps the most important ladder company duty in protecting exposures.

Interior Fire Extension

Ladder companies must check interior exposures to determine where the fire is located and to keep it from spreading to uninvolved areas. Fire can spread quickly within a building in almost any direction. Ladder company personnel must move just as quickly to find spreading fire and, where necessary, to open up building components to check the extent of the fire and to provide access so engine companies can extinguish the fire with their hose lines.

The protection of interior exposures must be a coordinated effort between ladder and engine companies. Ladder crews check for spreading fire, and engine crews extinguish it. Ladder company personnel should wait for a hose line to be advanced into the area before they begin to open up an area that is suspected to contain fire. They may be able to get into position more quickly than those on the hose lines to begin their firefighting tasks, but they should wait for hose line protection when encountering fire conditions. If fire is found in a particular location, hose lines will be necessary there. If not, the hose lines may be more useful elsewhere.

Fire fighters must check for fire extension in stairways and elevator shafts, through halls and corridors, and from room to room in involved areas. If fire is found to be spreading in these

> ## Key Points
>
> Ladder company personnel should wait for a hose line to be advanced into the area before they begin to open up an area that is suspected to contain fire.

> ## Key Points
>
> Fire fighters must check for fire extension in stairways and elevator shafts, through halls and corridors, and from room to room in involved areas. If fire is found to be spreading in these locations, there is a good chance that it is also spreading vertically and horizontally through concealed spaces.

locations, there is a good chance that it is also spreading vertically and horizontally through concealed spaces.

Fire in Concealed Spaces

There are signs to indicate that fire is spreading within a concealed space, but there are no signs to indicate that fire has not spread to a concealed space **Figure 6-2**. If there is any possibility of fire in a horizontal or vertical space or shaft, it must be opened and inspected visually. If necessary, streams must be directed into the vertical or horizontal passageway, and it must be ventilated.

Although this action will cause damage to the building, there is little choice in the matter. Either open shafts, walls, partitions, ceilings, floors, or other concealed spaces, or let the fire destroy the building. Although every effort should be made to minimize damage to the building and its contents, openings have to be large enough for inspection, hose line manipulation, and ventilation. Openings must also be large enough to admit a sufficient amount of water to extinguish the fire.

The fact that fire will spread vertically and then horizontally until blocked and that ventilation is important in controlling fire spread was established in the last chapter. It follows that ventilation and the search for fire in concealed channels are companion operations. The opening of concealed spaces and ventilation of involved areas are ladder company work that must be done by ladder company personnel in exposure protection operations.

Vertical Fire Spread

Fire will travel vertically inside walls and partitions and through utility shafts, dumbwaiters, air shafts, exhaust ducts, and similar pathways in a building. Many structures, including one- and two-family dwellings, contain concealed vertical shafts that carry water, gas and electrical lines, or sewer system vent pipes **Figure 6-3**. Many of the newer single-family dwellings and apartment houses have central-heating and air conditioning vents that extend up through the building to a chimney fixture on the roof.

These vertical channels are normally located toward the rear of commercial buildings, retail stores, and shopping centers, as shown in Figure 6-3a. In apartment buildings, they follow the pattern of the apartment layouts and are most often found near kitchens and bathrooms, each shaft usually located so it serves two, four, or more apartment units, as shown in Figure 6-3b. In some structures, such as an office building, all of the shafts are located in a single large central core that runs from the basement through the roof of the building, as shown in Figure 6-3c.

The great variety of designs for single-family dwellings means that vertical channels might be located almost anywhere

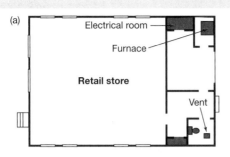

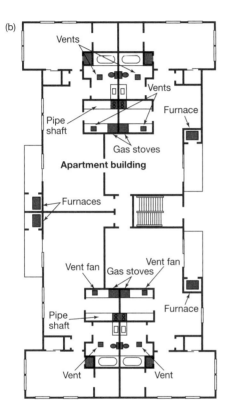

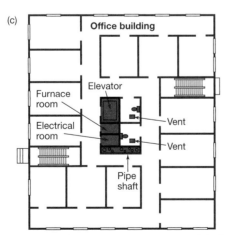

Figure 6-3 Locations of vertical channels vary with the occupancy type, size, and age of the structure.

in these homes. The locations of vent pipes and kitchen vents on the roof are good indicators of where these shafts will be found. Vertical concealed spaces are often created when the interior of a building is finished. Such spaces can be found under and around stairs and within walls and partitions. These openings will rapidly spread fire up through the building.

Indications If there is a working fire inside a building, fire fighters should assume that flames have entered concealed shafts until they determine otherwise. As they arrive on the fire ground, ladder company personnel should be looking for signs that fire has reached into vertical channels within the building. External signs, mainly involving roof features and the upper parts of a building, were covered in the previous chapter.

Inside the building, ladder crews should look for smoke and flames issuing from walls. Such signs as blistering, discoloration, or streaking of paint or other wall covering material indicate the presence of fire or heat within concealed shafts, walls, or partitions **Figure 6-4**. A wall that is hot to the touch may be concealing fire. At times, fire within a concealed space will actually crack, hiss, and pop loudly enough to be detected by its sound.

Key Points

A wall that is hot to the touch may be concealing fire. At times, fire within a concealed space will actually crack, hiss, and pop loudly enough to be detected by its sound.

Figure 6-4 When extending fire is found in walls or partitions, hose lines should be stretched and the wall opened. The area must be thoroughly checked.

Checking Walls   The search for vertical fire spread should begin directly over the fire on the floor above the fire floor.

Baseboard areas should be felt for heat and examined for black streaks running up the walls. If either is found, the baseboard should be removed so that the inside of the wall can be checked **Figure 6-5**. If fire is extending upward within the wall, a hose line should be called for and the wall opened to allow streams to hit the fire. If the fire is at the baseboard level only, the wall should be opened above the baseboard to check for fire extension. If there is heat only at the opened baseboard area and no fire, the wall need not be opened further.

The walls themselves should be checked for the signs of fire. A wall that shows any of these signs must be opened to

allow a stream of water to be directed onto the fire. The initial opening should be small and about waist high. With a hose line available, the opening should be enlarged so that the stream can be directed up into the wall. An opening at waist level also allows the stream to be directed down into the wall with comparative ease.

A wall opening should not be enlarged unless there is a charged hose line available with which to hit the fire. After the hose line is in position, the opening should be further enlarged until the extent of the fire is determined and the fire is knocked down.

When fire is found to be extending up past the wall opening, the area above it must also be checked. This is true whether it means checking another story, an attic, or a cockloft. If the fire has extended beyond that area, higher stories must be checked until the extent of the fire is found. This will require a coordinated effort in an expeditious manner by both ladder and engine crews alike. This operation is absolutely necessary. If this is not accomplished in a timely manner, the fire will extend into unburned areas with potentially disastrous results. Fire fighters must check for fire extension and keep ahead of the fire. Sufficient personal must be available to accomplish this task.

Checking a Vertical Shaft   Ladder crews on the floor above the fire must check all rooms that could contain utility shafts or pipes or any other vertical passageway. These include kitchens, bathrooms, workshop areas, laboratories, janitor's closets, and the like. Fire can spread vertically into these rooms through the shafts and other concealed spaces. In an apartment building, kitchens usually are located one above the other. Fire could enter

Figure 6-5   Check for vertical spread should begin directly over the fire. Baseboards can be removed to check extension through walls.

Key Points

Ladder crews on the floor above the fire must check all rooms that could contain utility shafts or pipes or any other vertical passageway. These include kitchens, bathrooms, workshop areas, laboratories, janitor's closets, and the like.

the kitchen above the fire floor through a utility shaft and then spread up from floor to floor through that same shaft.

Built-in cabinets below the kitchen sink usually are constructed with a 3- to 5-inch enclosed space between the floor and the bottom shelf. Electric conduits, gas pipes, water pipes, and drains often run through this enclosed space, and thus, it is open to a wall or shaft. Fire in the space will travel to the wall and then to higher stories if not located, exposed, and extinguished quickly. Fire entering the space from below will travel horizontally through the space, as well as vertically **Figure 6-6** . When over a fire, this area must be thoroughly checked.

Exhaust ducts in restaurants, hotels, hospitals, and other buildings with large kitchens are designed to remove smoke and grease-laden vapors. They often develop a heavy internal coating of grease. If ignited, grease burns with a very hot flame that can heat the duct to the point where it, in turn, ignites combustible material placed against or even close to it. If the seams of the exhaust duct are loose, flames can push through them and ignite parts of the building. Thus, whenever a fire involves a kitchen exhaust duct or a room in which a kitchen exhaust duct is located, the entire length of the duct should be checked, all of the way to the roof.

Ductwork for forced-air heating systems and central air conditioning systems will, over the years, become matted with

Key Points

Whenever a fire involves a grease duct or a room in which a grease duct is located, the entire length of the duct should be checked, all the way up to the roof.

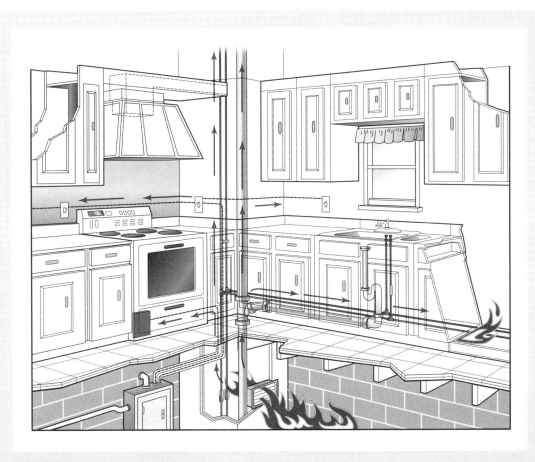

Figure 6-6 Kitchen cabinets often conceal vertical and horizontal shafts and openings, such as utilities for gas, plumbing, and electrical sources.

lint and dust. This material burns very fast once ignited, and the ducting system can quickly spread the fire throughout an entire building. Ladder crews checking floors above the fire must check air intake and outlet registers for smoke and the walls around ductwork for signs of heat. If either is found, a hole should be opened to determine the extent of the problem. If fire is found, a hose line should be brought into position, the wall opened, and the fire extinguished. The rooms above must be checked in the same way.

In some air conditioning systems, the spaces between pairs of wall studs are used as return air ducts. This type of setup is most prevalent in single-family dwellings and garden apartments. The stud-space "duct" sometimes is lined with thin sheet metal, but often is not. In either case, because it provides a channel for the spread of fire through the structure and into the walls themselves, stud-space ducts must be carefully checked for fire.

Pipe shafts might be completely concealed within walls or located behind service doors or wall louvers on each floor of a building. The concealed shafts are obviously hard to find, as can be shafts behind doors or louvers that could be anywhere in the building. Unfortunately, these shafts are sometimes located in the individual units of hotels, motels, or apartment buildings. In several notable cases, these shafts have contributed to loss of life when smoke, gases, and fire were not released at the roof in time to keep them from pushing their way into the living units.

When such shafts are located by fire fighters, they should be checked for signs of fire travel. If any of these signs is found, a stream of water should be directed up the shaft through a forced opening, an open service door, or an opening from which the louvers have been removed **Figure 6-7**.

When an intense fire is extending through a shaft, the floor and ceiling must be checked where they abut the shaft. If the floor is warm or hot or the ceiling shows any signs of fire, the area around the shaft must be opened and a stream of water directed into it. The quickest way to open the ceiling is to pull it down,

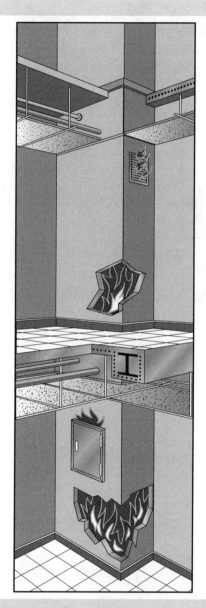

Figure 6-7   Pipe shafts, which might be completely concealed or located by noting the presence of service doors or louvers, must be found and checked when in a fire area.

starting with a small hole. If there is fire outside the shaft, the ceiling should be opened all around the shaft. As noted earlier, any shaft involved with fire should be opened at the roof.

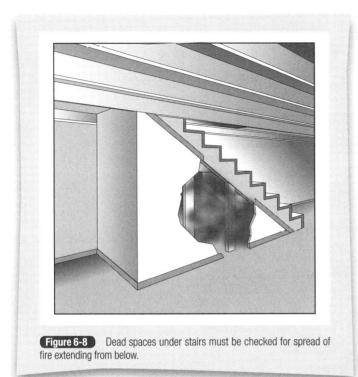

Figure 6-8 Dead spaces under stairs must be checked for spread of fire extending from below.

Checking Stairways Fire can start in or find its way into storage rooms or framed-out dead spaces under stairways. The fire could then quickly spread to adjoining walls or shafts or even into the ceiling by traveling along the underside of the stairway **Figure 6-8**.

Stairways and the spaces under them must be checked if they are near the fire, either on the fire floor or on the floor above the fire.

A fire may start in a room or area and extend out into the hallway or corridor. Fire can travel down a hallway to an open stairway where it could extend to the floor above. Because the stairways will be used by escaping occupants as well as by advancing fire fighters, if at all possible, the stairs must be kept intact and eliminated as a source of fire extension.

Checking Doors and Windows Carpenters often leave a space between a door or window frame and the adjacent wall studs. Shims are placed in the space to level the frame and steady it, but the effect is the same as that of a vertical shaft. Fire will quickly extend up around the door or window.

The areas around doors and windows should be carefully checked if they have come in contact with fire or if fire has burned into doors and windows on the floor below. If there is an indication

Key Points

Stairways and the spaces under them must be checked if they are near the fire, either on the fire floor or on the floor above the fire.

Key Points

The areas around doors and windows should be carefully checked if they have come in contact with fire or if fire has burned into doors and windows on the floor below.

that fire has extended into areas around doors and windows, any trim or framing should be removed and the area checked for fire.

Horizontal Fire Spread

Although the greatest tendency of fire is to travel vertically, it will also travel horizontally through any available paths. In the usual case, the fire starts in one area of a building and spreads to the ceiling, where it is temporarily blocked from further vertical spread by the ceiling and floor above. Eventually, it burns into the ceiling and walls and then starts to travel through the building. If vertical and horizontal channels are available, the fire will spread through both, but it will travel faster through the vertical channels. If it is blocked from vertical spread, it will travel quickly through horizontal channels.

Fire can travel horizontally through the spaces between ceilings and floors, over false or hanging ceilings, through a cockloft, through and along ductwork and utility conduits, through conveyor tunnels in factories and warehouses, and through similar channels in other buildings. In addition, fire can spread through the concealed horizontal channels formed within walls, floors, and ceilings by some types of construction. Suspended ceilings can be found in almost any type of occupancy, and they should be carefully checked. The space between the floor assembly and ceiling can be of varying heights and could contain various building systems, such as HVAC ductwork and electrical wiring and equipment **Figure 6-9**.

Fire also can move horizontally between attached buildings or occupancies and through ducts, ceiling spaces, and walls. An example would be the spread of fire from one adjoining apartment to another or through a row of stores. All horizontal channels must be checked for signs of extending fire.

Key Points

If vertical and horizontal channels are available, the fire will spread through both, but it will travel faster through the vertical channels. If it is blocked from vertical spread, it will travel quickly through horizontal channels.

Key Points

All horizontal channels must be checked for signs of extending fire.

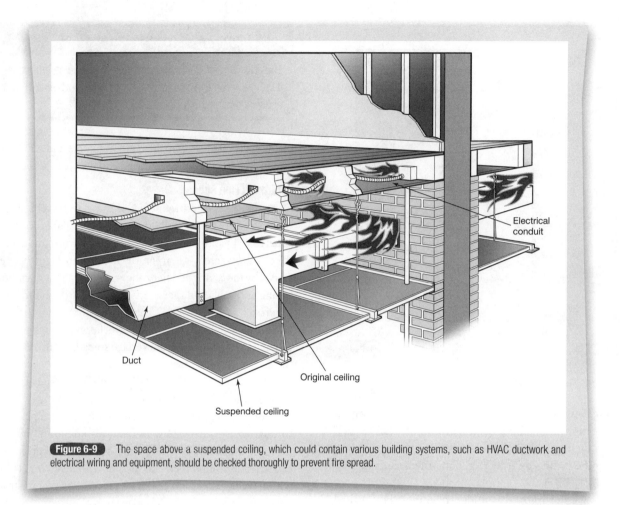

Figure 6-9   The space above a suspended ceiling, which could contain various building systems, such as HVAC ductwork and electrical wiring and equipment, should be checked thoroughly to prevent fire spread.

Indications   Ladder crews arriving at a fire building will see few external signs of horizontal spread unless the fire has reached and involved the exterior walls. Inside the building, the signs of horizontal fire spread are the same as the signs of vertical spread. Ladder crews must check floors and ceilings for smoke, fire, discoloration, hot spots, blistering paint, black heat streaks, and the sound of fire.

Checking Ceilings   When fire has control of an area, the ceilings of adjoining units such as in apartments, offices, and stores in the fire building, as well as those in attached buildings, should be opened to determine whether the fire is spreading. The fire is likely to spread horizontally from ceiling to ceiling.

Most ceilings are easy to open with a pike pole or plaster hook. First, a small slot should be opened along the common wall. If fire is found, a hose line should be called to extinguish the fire.

When it is in place, the ceiling should be opened up until the full extent of the fire is exposed and can be knocked down by the stream. If fire conditions permit, salvage covers should be placed over furniture, stock, or other valuables in the area to keep water damage to a minimum **Figure 6-10**; however, the most important objective is control of the fire.

Very high ceilings, such as those in older stores and in the lobbies of office buildings, are often difficult to reach and open with pike poles. Metal ceilings and those made of thick tongue-and-groove flooring material can take a long time to open. In such cases, it is probably best to open the flooring above the ceiling to check for horizontal fire spread.

On the other hand, ceilings consisting of square or rectangular tiles that lie on metal rails are very easy to open. The tiles are

Key Points

When fire has control of an area, the ceilings of adjoining units such as in apartments, offices, and stores in the fire building, as well as those in attached buildings, should be opened to determine whether the fire is spreading.

Key Points

Very high ceilings, such as those in older stores and in the lobbies of office buildings, are often difficult to reach and open with pike poles. Metal ceilings and those made of thick tongue-and-groove flooring material can take a long time to open. In such cases, it is probably best to open the flooring above the ceiling to check for horizontal fire spread.

Figure 6-10   A check for horizontal fire spread should be made on all sides of the main body of fire.

simply lifted off the rails with the tip of a pike pole or plaster hook. Where such tile ceilings are encountered, ladder crews must be quick to check adjoining areas for fire and smoke.

When older buildings are remodeled or when air conditioning or forced air heating is added to a building, a new ceiling is often constructed below the original ceiling. Thus, there may be two or even three levels of ceilings in such buildings **Figure 6-11** . After the lowest ceiling is opened, the others must also be opened—whether or not fire is found above the lowest ceiling. If necessary, the topmost ceiling can be checked by opening the flooring or roof above it.

In some mercantile properties, mainly supermarkets, super drugstores, discount stores, and stores in older buildings, hanging ceilings cover all or part of the total area. Hanging ceilings might be located above the entire sales floor but not the stockrooms

Key Points

When older buildings are remodeled or when air conditioning or forced air heating is added to a building, a new ceiling is often constructed below the original ceiling. Thus, there may be two or even three levels of ceilings in such buildings. After the lowest ceiling is opened, the others must also be opened—whether or not fire is found above the lowest ceiling.

or only above counters and display areas in which case they usually were installed to lower the lighting fixtures. Hanging ceilings, sometimes used for storing light goods and empty cartons, contribute to the rapid horizontal spread of fire across the building and must be checked carefully.

Checking Attached Structures   The cockloft or attics of structures attached to the fire building must be checked for the lateral spread of fire. Ladder crews must assume that there is nothing to stop the spread of fire through these spaces until they determine otherwise.

Proper ventilation will reduce lateral fire spread at the tops of exposed structures, as noted in Chapters 4 and 5, and venting crews on the roof might be able to detect fire spreading through attics; however, where there is any doubt, suspect areas should be checked through openings made in the ceiling below the attic or cockloft. In many fire situations, ladder crews will be checking an exposure for interior fire spread before venting crews are sent to its roof. In these cases, the attic or cockloft must be carefully examined by the fastest means.

The basements of structures attached to the fire structure, especially in older buildings, must also be checked quickly. The party wall between two attached structures might support the floor joists of both buildings; there is often an opening in the wall where the joists overlap. Deterioration of the mortar over the years could have created large openings through which fire can spread from one building to the other. In addition, large holes are often made in these party walls when plumbing or electric systems are modified or when new heating or air conditioning systems are installed. These holes are rarely closed up again, in spite of the fact that they will allow fire to spread into an uninvolved basement **Figure 6-12** .

Open Interior Spread

A fire that has gained control of a large open area, such as a supermarket, warehouse, or garage, can present engine companies with a serious problem. Ladder companies can assist in deterring the lateral spread of fire by closing doors, windows, service openings, and the like between the involved area and

Key Points

The cockloft or attics of structures attached to the fire building must be checked for the lateral spread of fire. Ladder crews must assume that there is nothing to stop the spread of fire through these spaces until they determine otherwise.

Key Points

The basements of structures attached to the fire structure, especially in older buildings, must also be checked quickly.

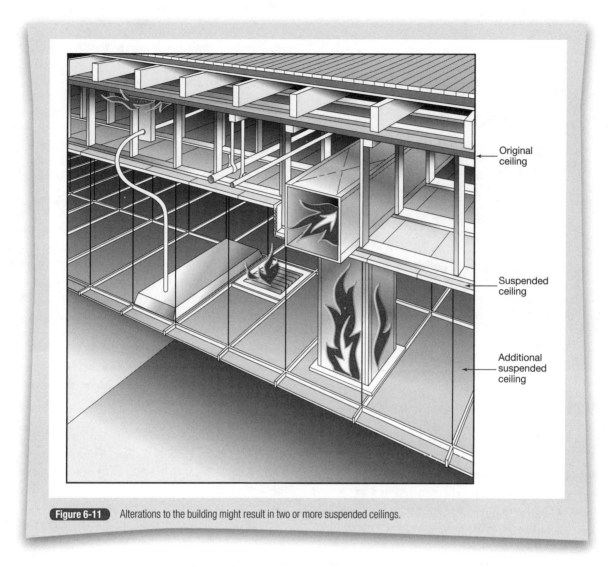

Figure 6-11 Alterations to the building might result in two or more suspended ceilings.

Original ceiling

Suspended ceiling

Additional suspended ceiling

the remainder of the building. This can often be accomplished while ladder crews are getting into position to check for fire spread.

Fire doors are sometimes blocked open by occupants to ease the flow of foot traffic. "Automatic" fire doors, which are normally open but which close in the event of a fire, are sometimes kept from closing by stored materials. Ladder company personnel should make sure these doors are closed properly. During inspections, building occupants should be warned about the dangers of interfering with the operation of fire doors. Generally, any opening between involved and uninvolved parts of the structure should be closed to slow the spread of fire.

Trench Cut

A **trench cut** is a procedure used to stop the spread of fire across the length of a fire building. It is used to set up a defensive line separating the burned area of the building from the unburned area, as shown in **Figure 6-13**. It is usually performed, but not limited to, buildings having somewhat of a narrow width such as a block of stores, a motel, or wings of buildings. It is usually performed on a flat roof but can also be done on a pitched roof.

Observation holes are cut in the roof to determine the location of the fire. The trench cut is then made well in advance of the observation holes. The cut, at least 4 feet wide, is made across the width of the roof from one exterior wall to the other.

There must be a concentrated effort using an adequate number of fire fighters with the proper hand and power tools to open the roof. This operation may be labor intensive and time consuming, depending on the construction of the roof and the length of time required to complete the cut. The trick here is to get the trench open before the fire gets to that location. Hand lines are used to keep the fire from advancing across the trench to the unburned area of the building. Fire fighters must be well supervised and use extreme caution when working on the roof during this operation.

Key Points

Any opening between involved and uninvolved parts of the structure should be closed to slow the spread of fire.

Figure 6-12 Fire fighters must check adjoining units for extension of fire if a basement fire is found in an attached structure.

Key Points

A trench cut is a procedure used to stop the spread of fire across the length of a fire building. It is used to set up a defensive line separating the burned area of the building from the unburned area.

Trench cut

Figure 6-13 A trench cut is made from bearing wall to bearing wall to prevent the horizontal spread of fire.

Exterior Exposures

Ladder crews can protect an exterior exposure from fire spread by entering the building and closing windows and outside doors to keep sparks, embers, and other burning material from entering. Fire fighters should remove curtains, drapes, and shades from windows that face the fire to keep them from being ignited by radiated heat. They should check air shafts and narrow walkways or alleys and possibly clear areas of any combustible materials that will prevent the fire from spreading toward the exposure. They also should evacuate all occupants of the building.

If the exposed building has a wet standpipe system, ladder crews could use the fire hose from the building often called the house line, if available, to knock down fire that has entered the building. This is not recommended for engine companies, who should use their own hose lines and nozzles, but ladder crews should use any available means to protect the exposed building.

If the main fire is not controlled or if exposure streams do not adequately protect the exposure, parts of the building could be ignited either by direct flame contact or by radiant heat. In this case, ladder crews must search for victims if they have not had the chance to evacuate the building. They also must inform command of the situation.

In general, ladder companies operating in an exposed building should take whatever action is necessary to slow the spread of fire into or through the building. After the exposure becomes involved with fire, it should be handled as a fire building rather than an exposure.

Chief Concepts

- Within a building, fire will travel through whatever paths are open to it, both vertical and horizontal. The major vertical paths are walls and vertical shafts and ducts. Horizontal paths include the space between a ceiling and the floor above, as well as horizontal shafts and ducts. These are hidden paths, which must be opened by ladder crews attempting to check for fire spread. When found, extending fire must be attacked by engine crews using appropriate-sized hose lines.
- Ladder crews can limit the spread of fire in large open areas and in exposed buildings by closing doors, windows, and other openings between involved and uninvolved areas. Some buildings, by virtue of their construction, are more prone than others to fire spread. Ladder company personnel should be aware of these buildings or this type of construction in their response area through preincident planning and inspections. This will allow them to operate more efficiently in checking for and controlling the spread of fire in these structures.

Key Term

Trench cut: A cut that is made from bearing wall to bearing wall to prevent horizontal fire spread in a building.

1. The major contribution of ladder companies to exposure protection activities is
   a. Checking for fire spread
   b. Operating master streams onto exposure buildings
   c. Operating master streams into the fire building from the exterior
   d. Operating master streams onto exposures or into the fire building depending on the volume of fire and radiant heat

2. A building attached to the fire building but separated by a fire wall would be classified as
   a. The fire building
   b. An external exposure
   c. An internal exposure
   d. Either an internal or external exposure depending on the degree of separation between the two buildings

3. The ladder company is checking an adjoining building and they find a concealed wall area between buildings that they suspect contains fire, they should
   a. Immediately open the wall
   b. Immediately open the wall and use a fire extinguisher to extinguish the fire
   c. Wait for a hose line to be advanced into the building before opening up the wall
   d. Wait for a hose line to be advanced to the area where they are working before opening the wall

4. If fire is found in adjoining _____, it is a sign that fire is spreading through concealed spaces.
   a. Stairways
   b. Halls and corridors
   c. Rooms
   d. Any of the above areas

5. When there is a working fire inside a building
   a. Only open walls if they are hot
   b. Open walls only if they are so hot that you cannot hold an ungloved hand on the inside wall
   c. Rely on the thermal imaging camera to determine whether there is a need to open a concealed space
   d. Assume that fire has entered concealed spaces

6. Exhaust ducts from large commercial kitchens are designed to
   a. Vent carbon monoxide and other combustion gases
   b. Provide air to the cooking surface when natural or liquefied gases are used for fuel
   c. Remove smoke and grease-laden vapors
   d. Depending on the type of fuel used, these ducts can be used for any of the above functions.

7. Fire has the greatest tendency to travel
   a. Vertically
   b. Horizontally
   c. Vertically or horizontally depending on the temperature of the fire, smoke, and gases being released
   d. In all directions

8. When encountering high ceilings made of thick tongue-and-groove flooring material that needs to be opened to check for fire extension, it is best
   a. To remove the boards one at a time from below using a pike pole
   b. Place an "A" ladder near the side walls and remove a board that is three or four board widths into the room using an electric or hand saw.
   c. Leave the ceiling closed, but check for fire extension high on the wall on the side nearest the fire.
   d. Go to the floor above and access the area between floors from there.

9. When making a trench cut, start by
   a. Making observation holes to determine the location of the fire
   b. Opening a hole from one exterior wall to the other across the shortest roof span
   c. Opening a trench in the approximate center of the roof
   d. Placing a hose stream in the attic to be protected

10. When making a trench cut, the width of the trench (roof opening) should be at least
    a. 1 ft   c. 4 ft
    b. 2 ft   d. 6 ft

11. When working in external exposure buildings, fire fighters should remove curtains, drapes, and shades from windows that face the fire to
    a. Provide a clear view of the fire building so lines can be operated from the exposure into the fire building
    b. Prevent ignition due to radiant heat
    c. Prevent ignition due to convected heat
    d. Prevent ignition due to conducted heat

12. If the exposure building has a wet standpipe equipped with a hose
    a. The ladder company should notify the engine company of the location of the standpipe connections within the building.
    b. The ladder company should request the engine company to pump into the fire department connection and have a fire fighter bring hose from the engine to connect to the standpipe.
    c. The ladder company should use the house line to knock down any fire that has entered the exposure building.
    d. The ladder company should use the house line to attack the fire in the fire building provided the fire is within reach of the hose stream.

Forcible Entry

Chapter 7

Learning Objectives

- Determine building characteristics, through preincident planning, that will assist the ladder company in gaining entry.

- Assess sizeup considerations and determine the best way for gaining entry to support other fireground activities.

- Identify and define common hand and power cutting tools.

- Identify and define common hand and power prying and forcing tools.

- Identify and define common striking tools.

- Identify and define common hand and power pushing/pulling tools.

- Examine procedures to provide entry through windows, doors, walls, and partitions.

Forcible-entry tools are used to gain entry into areas of a building that are locked or otherwise secured. Fire fighters must select the proper tool for the job so that entry into an area can be accomplished without delay.

Training is an important component for the safe operation of both hand and power tools used for gaining entry. Fire fighters must have a basic understanding of forcible-entry tools used for cutting, striking, prying, and spreading and know the safe limitations of each tool. Each member must be trained to operate each tool that has been selected to accomplish a particular task. They must adhere to established safety guidelines set forth by the department and should operate the equipment as the manufacturer recommends.

Fire fighters must be in full personal protective equipment (PPE) and are required to wear goggles or other approved eye protection when using forcible-entry equipment. All components of PPE will keep fire fighters safe from sharp cutting blades, sharp metal objects, and broken glass. They should know how to carry safely the tool to prevent injury to themselves or others. Tools should not be left on the ground when not being used, as they can cause injury to fire fighters stepping on or tripping over them.

Before swinging a tool or operating a piece of equipment, the area should be clear, and a fire fighter should be assigned to keep others out of the immediate area. Precautions should be taken when breaking glass to ensure that it will fall away from the hands and body. Fire fighters must be aware of their working environment at all times. A fire building can create a hostile atmosphere for those performing forcible-entry operations. Fire fighters must work safely while using the tools needed to accomplish the task.

To fight a building fire effectively, engine crews must advance their hose lines to the seat of the fire while ladder crews ventilate and check for fire extension, both in the fire building and in exposure buildings. Of utmost importance, fire fighters must perform a primary search for victims within these buildings. All of these operations require that fire fighters be able to enter the fire and exposure building quickly to perform these tasks.

One responsibility of ladder companies is to provide access to secured and protected structures, by force if necessary **Figure 7-1** . Ladder companies carry tools that can be used to cut, pry, and strike to force entry into a structure. They should carry tools designed especially for forcible-entry work. Ladder crews should know how and when to use these tools, as well as which buildings in their district may require forcible entry if they become involved with or exposed to fire.

This chapter deals with forcible-entry operations, from preincident planning and inspection to fireground techniques, but it is not the purpose of this chapter to teach manipulative skills. For that, there are a number of commercially available skill manuals, visual aids, and training manuals to supplement

Figure 7-1 Ladder company personnel should know how to use their tools in order to perform quick forcible-entry work.

the company's training sessions. Rather, this chapter deals with the efficient use of forcible-entry tools and skills in fire situations and with the effectiveness of forcible-entry operations.

Preincident Planning

The greater the ladder company's knowledge of its district, the more efficiently the crew will operate on the fire ground in forcing entry. As noted in previous chapters, preincident planning and inspections improve the performance of all fireground duties. In preparation for forcible-entry operations, ladder company personnel should determine

- Which buildings are secured during the day or night and may require forcible entry
- Which buildings are always open at the street entrance, but could require forced entry into individual units, such as in most apartment buildings, hotels and motels, and some office buildings
- Which buildings are locked at a street entrance and at an inner lobby door, both of which might have to be forced
- Which buildings have doors that, when locked, can be easily forced open, and those that are difficult to force
- Which buildings can be entered from the rear and/or the sides, as well as from the front. Normally, front entrances are the easiest to force, but the locations and construction of windows and doors at the sides and rear might allow them to be used most effectively for forced entry—especially if the front door is

difficult to force. Many times windows on upper floors are left opened, and entry can be gained by using a ground ladder.

- Which buildings have private security forces or maintenance personnel that will respond to an alarm with keys, thereby eliminating the need for forcible entry
- Which buildings have key box systems that contain all of the necessary keys to a building, as this box is usually mounted in a convenient location outside of the structure. Companies may also carry keys to a building in a box on the fire apparatus.
- Which buildings might present forcible-entry problems as exposures if a nearby or attached structure becomes involved with fire
- Which is the best way to enter problem buildings by force if that should become necessary

The results of such preincident planning and inspection can often be used in planning ladder company forcible-entry tasks. One example would be a case in which keys for the fire building were kept in a key box system; then it would not be necessary to assign fire fighters to forcible-entry duty on arrival.

Results of preincident planning might also be of use in pointing out the need for special forcible-entry tools in a particular district, in positioning ladder company tools on the apparatus so that the most used tools will be easiest to reach, and in assigning front and rear coverage and exposure coverage to first-alarm ladder companies.

Such preincident planning should be a continuing effort, as most building owners are constantly seeking to improve the security of their structures. As much as possible, ladder companies should be aware of changes in the way buildings within their district are secured. Changes made to increase the difficulty of unauthorized entry also increase the difficulty of fire fighters entering a fire building. Although it probably is impossible to know everything about a company's district, ladder crews should

> ### Key Points
> The greater the ladder company's knowledge of its district, the more efficiently the crew will operate on the fire ground in forcing entry.

> ### Key Points
> As much as possible, ladder companies should be aware of changes in the way buildings within their district are secured. Changes made to increase the difficulty of unauthorized entry also increase the difficulty of fire fighters entering a fire building. Although it probably is impossible to know everything about a company's district, ladder crews should certainly be aware of how best to enter those buildings with unusual or extremely difficult entry problems.

> ### Key Points
> Preincident planning, along with initial sizeup of the situation, will indicate the need for forcible entry. During sizeup, the type of occupancy, the life safety hazard, and the location and extent of the fire should be taken into consideration to support forcible-entry operations.

certainly be aware of how best to enter those buildings with unusual or extremely difficult entry problems.

Sizeup

Forcible entry is another ladder company operation that has a good possibility of adding damage to the fire building; however, the small amount of damage done through forcible entry—if the entry is accomplished properly—allows fire fighters to get into position quickly, can result in the saving of lives, and greatly reduces overall damage. Preincident planning, along with initial sizeup of the situation, will indicate the need for forcible entry. During sizeup, the type of occupancy, the life safety hazard, and the location and extent of the fire should be taken into consideration to support forcible-entry operations.

The Fire Building

First-arriving ladder companies might find little or no sign of fire, a working fire, or a smoldering fire in a building that must be entered forcibly. If there are no signs of fire, the building should be checked quickly to determine the easiest way to gain entry. In this situation, ladder crews have the time to force entry carefully so that they do the minimum amount of damage.

When a working fire has gained headway, especially when it threatens to cut off escape routes or has trapped occupants,

> ### Key Points
> If there are no signs of fire, the building should be checked quickly to determine the easiest way to gain entry. In this situation, ladder crews have the time to force entry carefully so that they do the minimum amount of damage.

> ### Key Points
> When a working fire has gained headway—especially when it threatens to cut off escape routes or has trapped occupants—arriving ladder companies must work quickly and decisively. They must force entry immediately without stopping to consider the damage they might do.

arriving ladder companies must work quickly and decisively. They must force entry immediately without stopping to consider the damage they might do. The faster the building is opened, the sooner the building can be searched, the fire attacked, and the products of combustion vented. A working fire justifies quick entry by the most expeditious means.

When arriving ladder companies find or suspect a smoldering fire, they must not enter the building until it has been properly ventilated.

Once inside the fire building, ladder company personnel might have to force entry to individual units such as apartments, offices, and commercial occupancies. Within these confines, there may be storage areas, utility and maintenance closets, mechanical spaces, and possibly fire doors. It may be necessary to enter any or all of these areas in order to perform other ladder company duties, especially to search for victims **Figure 7-2**. They also might have to force doors within the fire building to ensure that they will have access to units or areas of the building for later firefighting operations as needed. This will depend on the size and location of the fire, the type of occupancy, and the locations of locked doors relative to the fire.

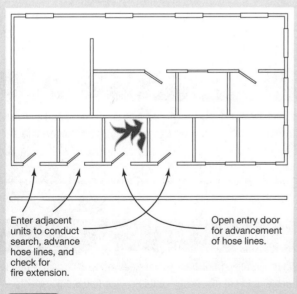

Enter adjacent units to conduct search, advance hose lines, and check for fire extension.

Open entry door for advancement of hose lines.

Figure 7-2   In apartments or office buildings, units on both sides of the fire and above the fire must be entered to conduct searches, advance hose lines, and check for fire extension.

Exposed Buildings

When the fire building is one of a row of attached structures, ladder crews should force entry into the locked buildings on each side of the fire building. They also might have to force entry into exposed structures that are taller than the fire building whether they are attached to the fire building or in close proximity, as shown in **Figure 7-3**. Whether or not ladder company personnel have to enter these buildings at the time they are opened, they should be opened so that fire fighters can quickly enter them if necessary.

The object of forcing entry into exposure buildings and exposed parts of the fire building is to provide access ahead of time. Fire fighters need to stay ahead of the fire so that no time will be wasted in entering the exposure if and when this action becomes necessary. Ladder company members must check to be sure they have provided access to all parts of the building. They might have to force inside corridor doors as well as the front door and perhaps even a lobby door.

Ladder companies should not force entry into exposures when force is not required. For example, if the exposure is a residential occupancy, the occupant, management, or security personnel might be available to assist in gaining entry. A quick check among bystanders for such a person might eliminate the need for forcible entry. In some cases, a cleaning crew working in an exposed office building can be quickly summoned to open the building. The entry door, or any door for that matter, should always be checked before it is forced because it may not

Figure 7-3 When detached buildings are exposed to the fire, their locked entrances should be forced so their interiors can be reached quickly if necessary.

have been locked. The saying "try before you pry" should be remembered while performing forcible-entry operations.

Tools

Forcible entry implies speed; a building is forced open so that fire fighters can enter it quickly. The forcible-entry operation itself should be carried out as quickly as possible and should create as little damage as possible. Both speed and minimal damage are achieved through proficiency with forcible-entry tools—a proficiency that comes only as a result of proper and continual training.

A ladder crew that is well trained in the use of forcible-entry tools will work quickly and efficiently with minimal damage. A poorly trained crew will work slowly and will not use the tools correctly. Incorrect use of the tools will result in excessive damage while slowing down other vital firefighting tasks.

Personal safety must be emphasized. Fire fighters can be injured using any type of forcible-entry tool. Special emphasis must be placed on the use of power tools, in training sessions as well as in actual operations. Eye and hand protection are of the utmost importance. Fire fighters must be in full PPE when performing forcible-entry tasks, as shown in **Figure 7-4**. A ladder company member who neglects this protection to save time will defeat the purpose if an injury results. Ladder company personnel performing forcible entry must be extremely careful when working with hand or power tools and air- or hydraulic-powered tools. Anyone not actually involved in the operation should stand clear of the area, allowing the crew member performing the task to do the work.

Key Points

Always check the entry door, or any door for that matter, before it is forced because it may not have been locked. Remember to "try before you pry."

Key Points

A ladder crew that is well trained in the use of forcible-entry tools will work quickly and efficiently with minimal damage.

Figure 7-4 Eye and hand protection must be worn when using any tools. If available, hearing protection should also be worn.

Cutting Tools

Cutting tools are designed to cut specific types of material. Cutting tools can either be operated by hand or by using power. Common hand cutting tools include axes, hatchets, saws, and bolt cutters. Power tools used for cutting include circular, chain, and reciprocating saws, hydraulic cutters, metal cutters, and oxyacetylene and exothermic cutting torches.

The cutting tools most often used for forcible entry are pick head and flat head axes, bolt cutters, power saws, and air-operated and hydraulic cutters. In addition, the adz (chisel) end of a Halligan or Kelly tool can be used to cut and can be driven by a flathead axe or maul. Of the two axes, the pick head is usually kept sharper for cutting. The flat head axe, not sharpened to as fine of an edge, is useful for forcing and prying.

Some departments use oxyacetylene cutting torches or hydraulic cutters for special entry problems. Torches are particularly effective in cutting bars away from windows and doors and in cutting through roll-up metal doors, but care must be exercised to make sure the torch does not start another fire.

Prying and Forcing Tools

Prying tools are designed to provide a mechanical advantage using leverage to force objects, in most cases, to open or break.

The hand tools available for prying and forcing work are the Halligan tool, claw tool, Kelly tool, Quik-bar, pry-axe, crow bar, claw tool, duckbill lockbreaker, large pipe wrench, and similar devices. Some of the common tools used for prying are shown in **Figure 7-5**. The flathead axe is often used for such work, either alone or driven by another flathead axe, or to drive some other tool **Figure 7-6**. For heavy work, a maul or sledgehammer can be used to drive another tool, or the maul can be used to drive the prying tool.

Hydraulic- and air-powered forcible-entry spreading tools are available in various sizes and types. Often such tools are purchased for nonfire rescue work, such as vehicle extrication, and then overlooked on the fire ground. The Rabbet tool is a hand-operated hydraulic tool used to pry open doors and windows. Windows and inward swinging doors are easily popped open with a few strokes of the tool's handle.

Depending on their design and capabilities, these power tools are used to force doors open, to raise roll-up doors, to remove or spread bars over windows and doors, and for similar

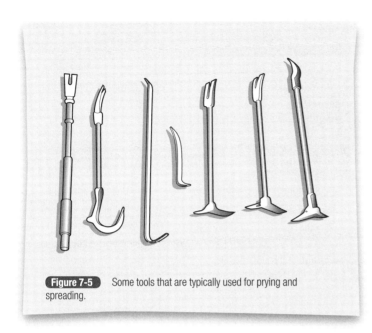

Figure 7-5 Some tools that are typically used for prying and spreading.

prying and forcing applications. In an area where heavy or barred doors are common, power tools should be standard ladder company equipment, and ladder company personnel should be well trained in their use.

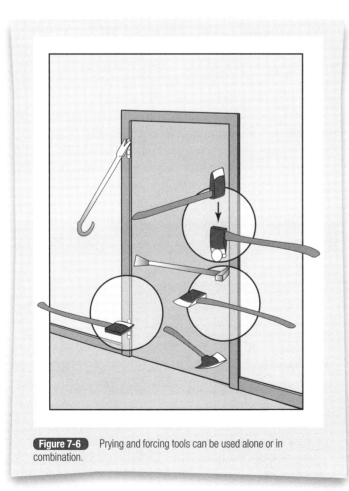

Figure 7-6 Prying and forcing tools can be used alone or in combination.

Key Points

In an area where heavy or barred doors are common, power tools should be standard ladder company equipment, and ladder company personnel should be well trained in their use.

Striking Tools

Striking tools are used in conjunction with other hand tools to perform different tasks on the fire ground, including forcible entry. Striking tools are usually used to drive another tool to provide leverage and a good bite on an object. Tools available for striking include the flathead axe, hammers, mallets, and mauls, sledgehammers, and battering rams. Some common striking tools are shown in **Figure 7-7**.

The battering ram has been shown to be effective for breaking through heavy doors and forcing openings through walls. Fire departments should equip each ladder truck with a battering ram.

Key Points

Striking tools are usually used to drive another tool to provide leverage and a good bite on an object. Tools available for striking include the flathead axe, hammers, mallets, and mauls, sledgehammers, and battering rams.

Key Points

The battering ram has been shown to be effective for breaking through heavy doors and forcing openings through walls.

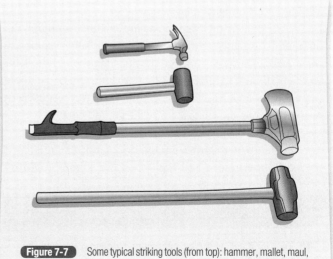

Figure 7-7 Some typical striking tools (from top): hammer, mallet, maul, and sledgehammer.

Key Points

A truck company with many stores, dwellings, and apartment buildings in its district should be equipped with at least one lock puller.

Lock Pullers

Lock pullers, such as the K-tool, are designed to remove cylinder locks. When operated properly, they do this quickly and are especially useful in opening wooden or metal doors equipped with cylinder locks. A ladder company with many stores, dwellings, and apartment buildings in its district should be equipped with at least one lock puller. In use, one part of the lock puller is driven onto the cylinder lock and then pried off with a Halligan bar or similar tool. The lock is pulled out with the tool and then a key tool is used to release the latch.

Pushing and Pulling Tools

These tools also have applications during forcible-entry tasks. Pike poles and plaster hooks in varying sizes can be used for breaking glass and opening windows. Inside a building, they can be used to break through partition walls and ceilings. Hydraulic rams can be used to spread doorframes apart and to pop open swinging doors.

Forcible Entry Through Windows

As noted earlier, the way in which ladder company members force entry to a building depends in part on the fire situation. If a working fire has control of a large part of the building or has trapped some occupants, entry must be forced the quickest way. This is usually through a ground-level door or window, which might have to be cut, pried, or forced open without regard for damage. It is usually much easier to force entry through a window than through a door.

When an arriving ladder company encounters no fire, the situation is usually one of an investigation. Ladder company personnel can make a quick examination of the building to look for the best place to force entry. In a situation where the doors and windows are locked on the ground floor, the best place to gain entry is often a window above the ground floor.

Thus, at its simplest, forcible entry may require only that one fire fighter climb a ground ladder, open an unlocked second-floor window, and then open the street door from

Key Points

It is usually much easier to force entry through a window than through a door.

Key Points

In a situation where the doors and windows are locked on the ground floor, the best place to gain entry is often a window above the ground floor.

Key Points

At its simplest, forcible entry may require only that one fire fighter climb a ground ladder, open an unlocked second-floor window, and then open the street door from inside. In contrast, the most serious forcible-entry problem might require cutting through a steel door or ramming through a wall.

inside. In contrast, the most serious forcible-entry problem might require cutting through a steel door or ramming through a wall. Forcible entry through windows is discussed in the remainder of this section and through doors and walls in the next section.

The way in which the forcible-entry operation is implemented also depends on the occupancy, especially the number of people who could still be in the building. This, in turn, usually depends on the time of day and the day of the week. The type of occupancy, its construction, and its population at the time of the alarm can reduce the options of ladder crews in forcing entrance to the building. Such information should be known before the alarm through preincident fire planning and inspection.

During a working fire, fire fighters usually gain access to a building through a ground-level door or window; however, street-level windows in some buildings are barred, shuttered, or otherwise fortified against burglary and vandalism. This is especially true of rear doors and windows in commercial

Key Points

The type of occupancy, its construction, and its population at the time of the alarm can reduce the options of ladder crews in forcing entrance to the building. Such information should be known before the alarm through preincident planning and inspection.

Key Points

Street-level windows in some buildings are barred, shuttered, or otherwise fortified against burglary and vandalism. When the fire situation requires that ladder crews force entry through such a window, they should use one above the ground level.

buildings. When the fire situation requires that ladder crews force entry through such a window, they should use one above the ground level. The window can usually be reached by a ground ladder or, if the building is so equipped, by a fire escape.

If a window at or above the ground level is unlocked, there is no entry problem. If the window is locked, it must be forced open. Depending on the situation, the quickest and easiest way to gain entry would be to break out the glass. If this is to be done, fire fighters as well as any civilians below the window should be notified and move away from the area to avoid shards of glass falling to the ground.

Double-Hung Windows

The window that allows the simplest and quickest access to a building is the large double-hung window. This type can be forced without much strain by prying up the bottom section at the center of the window. If the top section is made of small panes, the pane nearest the lock can be removed and the window unlocked, as shown in **Figure 7-8**. Metal-framed double-hung windows may present more of a problem than wooden-framed windows because of its inability to give while prying.

If a double-hung window that must be used for entry cannot be forced quickly, then the glass should be broken out of the window. If it is at ground level, this can be done with an axe or another appropriate tool. Fire fighters should carry appropriate tools when climbing a ladder to perform forcible entry. When they reach the window, proper precautions should be taken to ensure the safety of the fire fighter.

When a ladder is placed to a window for entry work, it should be positioned upwind if the wind is a factor. Any smoke, gases, or fire coming out of the window will be carried away from the fire fighters without endangering their lives or impairing their vision.

The fire fighter should take a position above the window and break the glass at the top. Hands should always be kept above

Key Points

When a ladder is placed to a window for entry work, it should be positioned upwind if the wind is a factor.

Figure 7-8 Double-hung windows can be opened by removing glass and unlocking or by prying.

Unlocking window

Prying window

the point of contact so that broken glass will fall down and away from the hands and body.

All glass should be removed from the window opening. This will avoid any injury to fire fighters or civilians who may pass through the window, as well as protect hose lines or any other equipment that may be taken through the opening. It reduces the possibility of anyone on the ground from being injured during the incident from falling shards of glass, and it will assist in ventilation.

In high-crime and vandalism areas, the glass panes of lower floor windows are sometimes replaced with unbreakable plastic panes such as Lexan or other plastic window coverings. Although this type of material can be cut with a circular power saw with a carbide tip blade, the quickest means of entry is to knock out the entire window frame if it is of wood construction. In some cases, knockout panels may have been installed in the window; these can be removed by striking a corner of each panel with the pick of an axe. If time is of the essence, another means of entry should be located.

The design of some windows prevents their use for quick access. They may have very heavy metal frames, wire within

the glass, horizontally hinged sections that swing out when the window is opened, center swing-out sections surrounded by stationary panes, and so on. A casement window is an example of a window that might be large enough for a fire fighter to fit through it, as shown in **Figure 7-9**.

Some windows may simply be too small to allow entry. Two types of windows that fall into this category are the projected or factory window and the Jalousie window. The projected window is heavy glass in metal frame attached on a pivot at the top or bottom. When this type of window is open, there is not much room for a fire fighter in full PPE, including SCBA, to maneuver and gain entry.

A Jalousie window is a series of heavy glass in small widths and lengths as long as the window opening. A wooden or metal frame is constructed around the window. A cranking mechanism is used to open and close the window. With the window open, there are only inches between the glass panels, which makes it impossible for a fire fighter to gain entry. Ladder companies should be aware of the locations of such windows in their district and should not expect to use them for forced entry. Alternative entryways should be determined in such cases.

Large windows, such as picture windows, of the double-pane type are expensive to replace. If such windows have been damaged by heat or smoke, there should be no hesitation in removing them; rather than forcing an undamaged double-pane window, however, some other quick and safe entry should be sought.

Figure 7-9 When a casement window is large enough to admit fire fighters, it should be opened by removing a pane, releasing the lock, and opening it.

Storm windows or screens must often be removed before built-in windows can be opened. In many cases, hooks or holding clips can be undone to allow these units to be removed quickly with little or no damage. A screwdriver could also be used to remove storm windows having such an installation; however, if this cannot be accomplished quickly at a working fire, the storm windows or screens should be broken to gain entry.

Barred Windows and Doors

For security reasons, many homeowners have chosen to install iron grating over windows. Storeowners have taken it a step further by installing rolling metal doors or barriers that, when closed, block the entrance to both doors and windows. This may tend to keep intruders from entering a building, but it also hinders fire fighters trying to gain entry into the building. Fire departments must have the proper tools and training to deal with this problem. Heavy metal bars and iron grating over windows may be difficult and time consuming to remove. Prying and striking tools may be used along with hydraulic cutters and/or spreaders and an oxyacetylene torch in an attempt to remove the security devices.

Rolling metal doors and barriers are usually secured with padlocks, which need to be removed. This security barrier should be opened fully to expose all windows and doors, which may need to be opened for entry or ventilation purposes.

Forcible Entry Through Doorways

When the fire situation demands that fire fighters quickly enter a secured building, ladder crews will probably be assigned to force entry through a doorway. Although there are a number of different types of doors, each type is more or less associated with a particular occupancy. For example, heavy metal doors can be found at the rear of a commercial structure, but rarely on

dwellings. Knowing the area or the particular building to which they are responding, ladder companies should therefore have some idea of the type of entry problem they might encounter. They should, however, possess the equipment and be prepared to force entry into every type of occupancy in their district.

Fire fighters should take into consideration the following items when determining what method to use to gain entry through a particular door:

- How is the door constructed
- What is its use
- How is the door hung
- How is the door locked
- The proper tools available to force the door open

Commercial Occupancies: Front

It is almost always easier to force entry through the front door than through the rear door of a store or other business establishment. In an older building, the front door might be constructed entirely of wood or of a wood frame surrounding ordinary plate glass. In more modern structures, the front door is often made of tempered glass, with or without a frame, or of heavy plate glass in a strong metal frame. One, two, or more such doors can be set into the doorway. Rear doors are usually made of steel or reinforced with steel.

The front doors and display windows of many business occupancies are further protected by metal shutters, accordion-type barred gratings, or similar devices installed to prevent vandalism and burglary; however, these devices also hinder firefighting operations. Such a device must be forced open before the door or display window can be opened. Most of these protective devices are secured with a padlock that can be forced open with a Halligan or claw tool. Others, locked into a device set in concrete, take longer to force. Whichever way they are locked, the devices must be opened so that fire fighters can get to the doorway.

Tempered-Glass Doors   Tempered glass is much stronger than plate glass because of the procedure of heat treatment during tempering in the manufacturing process. It increases its resistant

Key Points

Tempered glass should be considered as a last resort for gaining entry into a building.

to shock, impacts, and temperatures, making tempered glass ideal for office, institutional, commercial, and manufacturing occupancies.

Tempered glass should be considered as a last resort for gaining entry into a building. Another means of entry should be located before attempting to break out a doorway with tempered glass. If a tempered door must be used to gain entry, it should be attempted at its lock.

The locks on tempered-glass doors are usually cylinder locks, located at the middle or at the bottom of the door. Double tempered-glass doors are usually locked at the middle. In either case, a lock puller can be used to remove the lock. When no lock puller is available, the adz end of a Halligan, Kelly, or similar tool can be used. If the lock is at the middle of the door, the tool is driven in between the door and frame or between the two doors of a double door until the lock is forced open. In an alternative procedure, the tool is first driven into the space above the lock and then driven down to destroy the locking pins. If the lock is at the bottom of the door, the claw end of the tool is driven under the door until the lock is raised out of its keeper. This method will work with any type of lock located at the bottom of a door. Also, hydraulic tools such as a small porta-power or Rabbet tool can be used to force apart double doors or to raise doors locked at the bottom. Precautions should be taken noting that tempered glass is almost impossible to flex because of its rigid characteristics. If the door is pushed too far out of shape, the glass could break. It will disintegrate into hundreds of small pieces. Fire fighters should use full PPE and have eye protection in use when working on a tempered-glass door.

In many front entrances, tempered-glass doors are set between stationary plate-glass panels. The panels are usually as high as the door and wide enough to admit fire fighters. The quickest way to force entry through such a doorway is to break out the plate glass, leaving the tempered-glass doors in place **Figure 7-10** .

Many stores, especially larger ones, have one or more display windows near the tempered-glass entrance doors. Again, the doors should be left in place and the building entered by breaking

Key Points

Fire fighters should use full PPE and have eye protection in use when working on a tempered-glass door.

Figure 7-10 When locked tempered-glass doors are close to plate-glass windows, enter through these windows. If entry must be made through the door, force or pull the lock.

Key Points

In many front entrances, tempered-glass doors are set between stationary plate-glass panels. The panels are usually as high as the door and wide enough to admit fire fighters. The quickest way to force entry through such a doorway is to break out the plate glass, leaving the tempered-glass doors in place.

the glass in the display windows. This is especially effective where the display windows open directly to the sales floor; then they can be used for advancing attack lines, venting, removing victims, and other firefighting operations.

If a tempered plate-glass door must be broken, it is best to do so by striking a lower corner of the door with the pick end of an axe. As noted previously, when the door breaks, it will shatter into hundreds of pieces, not sharp, but more like globules; however, the fire fighter attempting to force the door should position as much of his or her back toward the door as possible when using the axe. Other fire fighters should stand back from the door and the axe.

Heavy Plate-Glass Doors Doors with heavy plate glass set in heavy frames should be treated in the same way as tempered-

Key Points

Many stores, especially larger ones, have one or more display windows near the tempered-glass entrance doors. Again, the doors should be left in place and the building entered by knocking out the glass in the display windows.

glass doors. Although it might be possible to remove the glass quickly, there is usually a bar across the center or lower center of the door that will need to be removed before fire fighters could easily get through the opening. In addition, the glass could shatter and be thrown at fire fighters if the door is pushed too far out of shape while being forced. It is much better to remove or force the lock or to enter through a nearby plate-glass window.

Wooden Doors Wooden doors leading to the outside are usually of solid core construction, meaning that the entire door or the core of the door is made of a solid material. Residential doors usually open into the building, whereas doors used in public buildings usually open to the outside so that people can exit quickly in an emergency.

Wooden doors may or may not have cylinder locks, but they usually have bolts that engage keepers at the doorjamb. This bolt may be an integral part of the lock assembly or it may be a separate lock. Wooden doors may also have a lock at the top or bottom of the doorway or at both places. Double wooden doors can be bolted to each other, as shown in **Figure 7-11**, and thus, pulling or forcing the lock does not guarantee entry; however, these doors usually have center panes or panels that can be broken out quickly. A fire fighter can reach through the opening and unlock or unbolt the door from the inside.

Commercial Occupancies: Rear

Commercial occupancies usually can be entered at the front. Metal doors, barred doors and windows, and roll-up doors very often complicate rear entry. There are seldom any display windows at the rear; however, where a rear door is constructed like a front door, it should be forced in the same way.

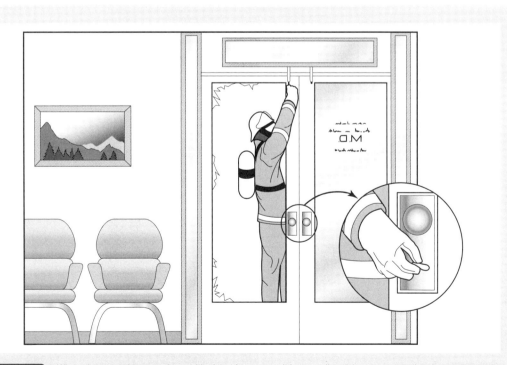

Figure 7-11 Where there are older type doors with plate-glass panes, it is generally quickest to remove the glass and open the door from the inside.

Metal Doors Metal doors are usually not solid metal but metal covered, hollow core, or constructed with some type of filler sandwiched in between. These doors may be difficult to force open if the metal frame is set in concrete, brick, or masonry.

Before an attempt is made to force a metal door, it should be checked for an exposed lock or exposed hinges. If the lock can be seen, drive in a pry tool between the door and frame, and force the door open. If the hinges are exposed, pull the hinge pins and open the door from the hinge side, as shown in Figure 7-12 , or drive a tool between each hinge and the door facing to force the hinges loose.

Doors with neither a lock nor hinges exposed cannot be forced with standard tools. Such doors might be secured with a steel bar placed across the door and doorframe and held in place by heavy steel hangers. They might be locked with a **fox lock**—a device with from two to eight bars that hold the door closed from the inside. In a fox lock, the bars are attached to a rotating plate on the door. The plate is rotated one way to move the bars into keepers in the doorframe, and the opposite way to withdraw the bars and allow the door to be opened.

A metal door that cannot be forced can be cut open or its lock cut out with a power saw or a cutting torch. An attempt could be made to open a heavy metal door with a battering ram. The ramming may force the locks from their keepers, bend the door enough to pull the locks from their keepers, or tear the keepers out of the wall or doorframe.

A metal door equipped with a fox lock is practically impossible to force. The operation is so time consuming and requires so much effort that by the time the door is finally opened, it is usually no longer effective for firefighting operations. If possible, an alternative entry should be found.

Fire fighters should remember to check all doors to determine whether they are unlocked before forcible entry is attempted. Check for the location of the hinges that will determine if the door opens inward or outward. Look for another means of

Key Points

Before an attempt is made to force a metal door, it should be checked for an exposed lock or exposed hinges.

Key Points

A metal steel door that cannot be forced can be cut open or its lock cut out with a power saw or a cutting torch. An attempt could be made to open a heavy metal steel door with a battering ram.

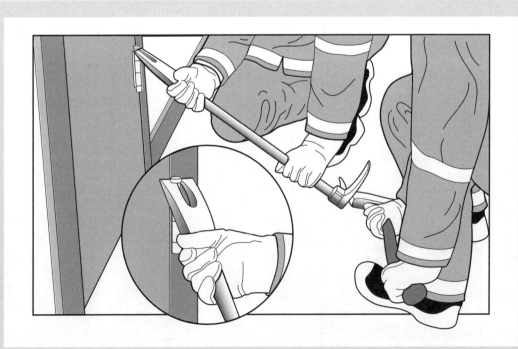

Figure 7-12 When hinges are exposed on metal doors, the quickest way to gain entry is usually to remove the pins and force the door from the hinge side.

gaining entry if a particular door presents a difficult forcible-entry problem.

Overhead Doors   Overhead doors are usually constructed of wood, wood with glass panes, metal with or without windows, and fiberglass. Overhead doors can be further categorized as sectional or folding, rolling, or slab.

Doors that open upward might be locked in any one of several ways. On an overhead door, the lock and a handle are usually located at the center of the door, and the bars engage keepers at each side. On a sectional or folding overhead door, a door panel near the lock can be knocked out and the door opened by reaching in to rotate the lock handle.

A wooden overhead door might be further secured with pins that extend from the sides of the door into the track on which the door rides. The door should be pried up from the bottom to bend the pins out of place. In some cases, a ring on the door is padlocked to a ring set into the floor. This arrangement can be opened by forcing the claw end of a Halligan or claw tool under the door against the rings and driving the tool with a flathead axe or sledge. **Figure 7-13** shows the various ways to gain entry through an overhead door depending on the type of door it is.

If a wooden overhead door is difficult to force open, it can be cut with a power saw or axes. When cutting or prying these doors, ladder crews must be careful of glass panes set in the doors. If the glass breaks while a door is being forced, shards of glass may fall in all directions.

If an overhead door is secured with a padlock (which could be located at either side lock or on the track itself), a fire fighter will have to gain entry through a panel or other opening cut in the door to cut the lock so that the door can be raised. Slab doors are opened in the same manner as a sectional or folding overhead door.

Metal overhead doors do not usually have built-in locks. They can be padlocked to the floor or locked into their rails with pins. Manually operated doors are often locked through the chain used to raise and lower them. A motorized door is rigidly connected to its operating mechanism.

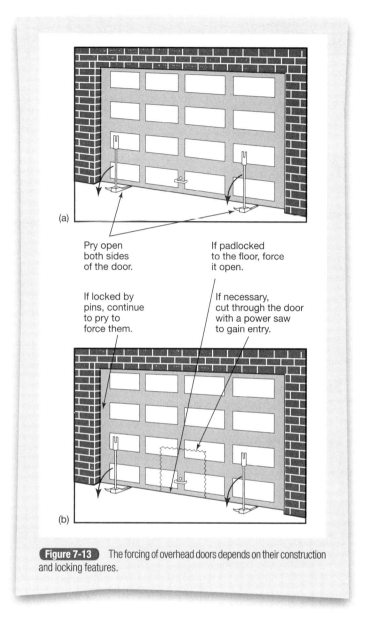

Figure 7-13   The forcing of overhead doors depends on their construction and locking features.

The first step in forcing a metal overhead door is to pry it up as much as possible at both sides. If the door is locked to the floor with rings and a padlock, the locking assembly should be forced as described earlier. If no locking rings are found, the door might be locked with pins or through its chain. Continued prying may bend the locking pins out of the rails or separate the chain. A hydraulic spreader can be of help here.

If only one side of the door can be raised, it might be possible to provide adequate access by wedging up that side. In some cases, one fire fighter might be able to get inside, under the door, and then release the latches to get the door opened. If nothing seems to work and the door must be opened, a hole should be cut in the door with a power saw or oxyacetylene cutting torch.

Light Doors   In many older buildings, rear doors are made of wood or light metal, reinforced with bars, or fitted with several locks. This is often the case in older row structures with small shops on the ground floor and apartments or offices above.

The main lock of such a door should be forced first. If it can be sprung, the additional bolts and locks usually can be forced with hand tools and brute strength. The various pry tools and striking tools such as the flathead axe will be most effective; the axe can be used to drive other tools or to pry the door.

If the door has a glass pane without bars, it is best to remove the glass and attempt to open the locks from the inside provided the location of the fire does not prevent such action. As always, fire fighters must use caution when attempting to force a door that contains glass.

Dwellings and Apartments

Locked residential structures are, in general, more easily entered than commercial structures. Front and rear doors are usually of the same type and of light construction. They often have one or more glass panes that can be broken, allowing ladder crews to open the locks from inside.

The hand tools carried on the apparatus are more than adequate to quickly force entrance into a locked one- or two-family dwelling. Ladder company personnel may need to force entry into a multiple-unit residence because the street doors at the front are often locked for security reasons. The rear doors might also be locked during the day as well as at night. There may be a lobby door, secured by an electric lock that can be opened by request from each apartment. Residents or building representatives may be present or a key box system may be

in place, allowing fire fighters to enter the building without forced entry. In older buildings, entrance doors are light, usually of wood and glass, and can be forced easily. Newer apartment buildings might have tempered-glass lobby doors, with plate glass at each side. These doors usually can be forced at the electric lock, or the plate glass can be broken.

In modern high-rise and garden apartment buildings, the lobby doors and the doors to interior stairways are usually unlocked and/or a building representative may be stationed in the lobby; however, if the doors are locked, they can be forced quickly by methods described earlier in this section.

Apartment Doors   Once inside an apartment building, ladder crews might have to open individual apartment doors to perform the primary search, ventilate, and check for fire extension. In older buildings, apartment doors are usually made of wood; cylinder locks might have been added and the original door locks retained. Cylinder locks can be pulled with a lock puller and the doors forced with any prying tool. The frames of these doors are usually strong enough to support a pry tool, allowing doors to be forced fairly easily.

In more modern apartment buildings, apartment doors are made of metal or of wood covered with metal. Most often, they are secured with cylinder locks and possibly one or more bolt-type locks. Here again, a lock puller should be used to begin the forced entry, or a handheld pry tool should be placed just above or just below the lock, driven in with a flathead axe, and then grasped at the outer end and pulled away from the door. Use of the porta-power or Rabbet tool is also effective for quick opening of these doors. The pressure exerted in forcibly opening an apartment door will usually tear loose or break any chain lock.

Floor Locks (Police Locks)   In high-crime areas, ladder companies might encounter the floor lock, or brace, as shown

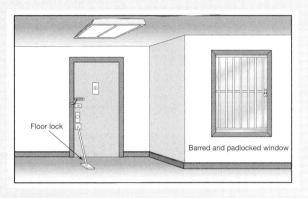

Figure 7-14 When fire fighters encounter a door with a floor lock, entry might be accomplished more rapidly by breaching the wall from the corridor or from an adjoining unit if it can be entered quickly.

in **Figure 7-14**. This device consists of a heavy bar fastened to the floor inside the apartment door and to a plate on the door. The bar can be pivoted away from the door when the tenant opens the door, and it slides into a plate mounted on the door for locking. Where floor locks are in place, entry can be very difficult, especially if metal doors are used.

A floor lock is very difficult to force. A wooden door can sometimes be pulled away from the device at the hinge side, or it might be possible to cut the door open with an axe, allowing someone to reach in and release the floor lock; however, in some of these devices, the bar is locked to the plate in the door, and the door must literally be destroyed before the apartment can be entered.

The combination of a floor lock and a metal-covered wooden door presents an even greater problem. It is extremely difficult to break through the door or to open it at the hinge side. To add to the problem, a police lock is usually used in combination with barred windows. If this is the case, the simplest course may be to breach a wall to get into the apartment. The hallway wall should be opened quickly and the door left in place if it is known or suspected that occupants are in the apartment. If an adjoining apartment has been opened or can be entered quickly, it might be easier to breach the wall between the two units. Common apartment walls are usually of lighter construction than corridor walls. Entry through a window, balcony, or porch by ground ladder or aerial device may be another option.

Sliding Doors In many residential dwellings and modern apartment buildings, sliding glass doors lead from a room to a deck or yard in dwellings or in individual apartments to balconies. When located in close proximity to the fire situation, these doors can be used for entry and fire attack. The doors might be equipped with cylinder locks or with some bolting arrangement holding them at the top and bottom. The framing around the doors is usually aluminum or another light metal.

A cylinder lock can be pulled or forced as previously discussed. The locations of bolts can be determined by prying

with a Halligan or claw tool or an axe. After located, the bolts should be forced with the available tools. Usually this will spring the doors away from the light framing.

If a door is particularly tough to force, use a flathead axe to drive a pry tool between the door and the framing. Two sliding doors locked to each other also can be opened by driving a pry tool between them or by using a porta-power or rabbit tool, as shown in **Figure 7-15**.

Because there is so much glass involved, care should be taken to avoid straining the glass enough to break it. As a rule, the glass should be broken out for entry when there is a need for immediate rescue or when the glass is already stained or damaged by heat or smoke; however, when a bar or rod holds the sliding section of a door in place for security reasons, the glass will have to be broken to reach and unlock the door or remove the bar.

Office Buildings

Generally, forcible entry into the units of an office building presents the same problems as entering the units of an apartment building.

Figure 7-15 A rabbit tool—a small hydraulic spreader—is useful in opening doors.

Key Points

As a rule, the glass should be broken out for entry when there is a need for immediate rescue or when the glass is already stained or damaged by heat or smoke.

Key Points

Generally, forcible entry into the units of an office building presents the same problems as entering the units of an apartment house.

The age of the structure usually determines the type of inside office door, unless the building has been remodeled extensively.

Most office buildings are open to the street during the day but locked after business hours and on weekends. Security and maintenance people could be working in a building while it is closed, but they might not be near the entrances when fire companies arrive.

The type of door and the material around it determine how entry will be made. Outside entrances to office buildings are usually similar to those found in stores of the same general age. Modern buildings have tempered-glass doors, and the older buildings use light metal or wooden doors.

Warehouses and Factories

These structures usually have overhead doors at loading platforms and heavy wooden or steel pedestrian doors. In addition, the windows on lower floors may be barred. These types of doors have already been discussed, as have window bars that can be pried apart with hydraulic spreaders or hand pry tools.

Fences   Many warehouses and factories are surrounded by fences, which are usually of the chain-link type. After working hours, fire fighters might have to force their way through the fence before they can get to the fire building. This may only require cutting a chain with a pair of bolt cutters or forcing a padlock or two with a Halligan, claw tool, or bolt cutters so that a gate can be opened. If padlocks on the inside of the gate are difficult to reach, however, the fence can be bridged with ground ladders (Chapter 9) and a fire fighter sent inside the fence to force the locks.

Usually, one gate is padlocked from the outside. Its location should be known through preincident planning and inspection. First-alarm response routes should be set up so that fire companies arrive at that gate. Where possible, fire companies should use a key box system or arrange to have on-duty plant security personnel open the gates and accompany fire fighters to the involved building in the event of a fire. If a chain link fence needs to be cut, a pair of bolt cutters should be used to perform the task.

Key Points

Many warehouses and factories are surrounded by fences, which are usually of the chain-link type. After working hours, fire fighters might have to force their way through the fence before they can get to the fire building.

Key Points

Usually one gate is padlocked from the outside. Its location should be known through preincident planning and inspection. First-alarm response routes should be set up so that fire companies arrive at that gate.

Guard Dogs   A word of caution: In some areas, these occupancies are protected at night by guard dogs roaming inside the fence. The fire department should know, through preincident planning and inspection, where guard dogs are being used; in some communities, ordinances require that the fire department be notified of such protection and the building posted. A system should be set up to ensure that the agency supplying the dogs is notified as soon as possible after first-alarm companies are dispatched.

Combination Occupancies

So-called loft buildings, with several types of occupancies, almost always present a double-entry problem. After business hours, entry may need to be forced first into the building itself and then into the individual occupancies. The doors installed in such buildings are similar to those already discussed; those leading to individual units can be heavily barred and bolted.

Sidewalk Basement Entrances

Doors leading to a basement, either from outside a building or from inside on the ground floor, are usually constructed like those already discussed. The one exception is the sidewalk entrance, consisting of two metal doors installed flat, or nearly so, on the sidewalk. The doors open upward to provide a large opening into the basement, which can be used to advantage in many firefighting situations.

Sidewalk basement doors are either manual or automatic. The manual type can open onto a stairway leading down or can only provide an opening to the basement. The automatic type is rigged to operate with an electrically driven elevator; the doors

Key Points

So-called loft buildings, with several types of occupancies, almost always present a double-entry problem. After business hours, entry may need to be forced first into the building itself and then into the individual occupancies.

open as the elevator rises to the street level. A bell and a sign are mounted on the building near every set of automatic sidewalk doors; the sign indicates that the doors are automatic, and the bell rings when the doors are opening.

Sidewalk doors are sometimes difficult to open, especially the automatic type. They are rarely padlocked from above because the padlock would interfere with pedestrian traffic, but instead, they are usually locked from below with a bolt or sliding latch. If a padlock should be found on a set of sidewalk doors, however, it can be quickly forced with a Halligan or claw tool.

Most often, the forcible-entry operation is more difficult. The location of the lock, on the underside of the doors, will be indicated by several bolts or rivets somewhere near the overlapping part of the doors. The doors should be pried apart as much as possible at that location so that the lock can be seen. If the lock is of the bolt type, a tool should be placed against it and driven in with a flathead axe. This will tear the lock loose, but it may be a tough job. If the lock is of the latch type that swings parallel to the doors, it might be possible to get a tool onto the swing latch to drive it out of its keeper.

When the lock cannot be forced, the hinges of one door should be attacked. After these hinges are broken loose, both doors can be lifted off the opening as a single unit. It may be possible to drive the hinges loose with a maul or to break them off with a Halligan, Kelly, or claw tool driven with a maul or a flathead axe.

If necessary, a power saw can be used to cut through the door to the lock. Cuts should be made around the lock so that the door sheeting can be pried up and the bolt or latch released. Knowing the type of lock involved through preincident planning and inspection increases the effectiveness of the forcible-entry operation.

Breaching Exterior Masonry Walls

It is sometimes quicker to open an exterior wall than to open a reinforced metal door, especially if the wall is made of concrete block or cinder block. Mauls, battering rams, and power tools such as a circular saw with a masonry blade, air chisels, and hydraulic spreaders can be used to make openings in most walls, including brick. An aggressive attack with proper tools often will open a wall in a reasonable length of time.

If possible, the wall should be opened near the doorway. The door and doorframe will help support the opened wall, and the opening should lead to a corridor or other open area. If a fire condition exists inside the area, the opening should be made only large enough to permit streams to be directed inside to knock down fire and cool the interior. If the opening is to be used to gain entry, the hole must be enlarged enough to allow access for fire fighters. At this point, ladder crews must be sure that blocks or bricks over the opening are firmly in place.

Breaching Interior Partition Walls

It is sometimes easier to open an interior wall than to open a door inside the building. It may also be necessary to open an interior wall to pass into another area when a door is not available for that purpose. Construction materials used to partition off areas of a building may not be readily known to fire fighters. They may have to make an opening to determine what is behind the wall. Concrete block walls and hollow clay tile should be opened following the same procedures for opening exterior masonry walls. Sheet rock and dry wall can easily be opened with cutting and prying tools as well as pike poles and plaster hooks. Cut along the studs and make the opening wide enough to pass through. Electrical and/or other utilities should be checked before and during the breaching of the wall. Plaster and lath may be found in older occupancies and may be breached in the same manner as dry wall.

Wrap-Up

Chief Concepts

- Preincident planning and inspection are important parts of forcible-entry operations. Ladder companies must be familiar with the different types of entrances to buildings in their district and with the tools needed to force open these entrances. Fire fighters also should be aware of buildings that would present especially difficult entry problems if they became involved with or exposed to fire. Ladder companies need training to cope with these problems, using specially designed forcible-entry tools if necessary.

- Depending on the fire situation, it might be easiest and fastest to force entry into a building through windows. Above the first floor, consider entry through a window, balcony, or porch by ground ladder or aerial device. Otherwise, doors must be forced or walls must be breached. The most difficult doors to force are those that are made of tempered glass or reinforced metal. Thus, it is important that the apparatus carry a full range of forcible-entry tools. These should include standard hand tools for cutting, prying, and striking to power saws, hydraulic cutting and spreading tools, air-powered cutting tools, and chisels and oxyacetylene torches.

Key Terms

Cutting tools: Tools that are designed to cut into metal or wood.

Fox lock: A device with from two to eight bars that hold the door closed from the inside.

Lock pullers: Such as the K-tool, are designed to remove cylinder locks.

Prying tools: Tools designed to provide a mechanical advantage using leverage to force objects, in most cases, to open or break.

Striking tools: Tools designed to strike other tools or objects such as walls, doors, or floors.

1. Forcible-entry operations can
   a. Add damage to the fire building
   b. Allow fire fighters to get into position quickly to reduce overall damage
   c. Allow fire fighters to get into position quickly to save lives
   d. All of the above

2. When first-arriving ladder companies are confronted with a working fire that has gained headway with a possible life hazard, they should
   a. Perform a quick 360-degree walk around, checking for alternative access to the fire building that does not require forcible entry, heeding the adage "try before you pry."
   b. Quickly check the fire building to determine the easiest way to gain entry and limit damage.
   c. Force entry into the fire building immediately without stopping to consider the damage.
   d. Force entry into the fire building immediately, but consider the damage.

3. When confronted with a working structure fire with exposures on either side, ladder crews assigned to forcible entry should
   a. Check the door to be sure it needs to be pried and also check with management, occupants, or security to see whether they can open the door, heeding the adage "try before you pry."
   b. Quickly check the exposure building by performing a quick 360-degree walk around all involved buildings to determine the easiest way to gain entry and limiting damage.
   c. Force entry into the exposure buildings immediately without stopping to consider the damage.
   d. Force entry into the exposure buildings immediately, but consider the damage.

4. The _____ axe head is not usually sharpened to as fine an edge as other axes.
   a. Flat   c. Pick
   b. Chisel   d. Adz

5. The Rabbet tool is used to pry open doors and windows, but not
   a. Inward swinging doors
   b. Outward swinging doors
   c. Metal fire doors
   d. Metal-framed double-hung windows

6. It is usually easier to force entry through a
   a. Metal door
   b. Metal-clad wooden door
   c. Window

7. When breaking window glass from an upper story window, the fire fighter should
   a. Work from the upwind side and above the window
   b. Work from the upwind side and alongside the window
   c. Work from the downwind side and above the window
   d. Work from the downwind side and alongside the window

8. It is almost always easier to force entry through the _____ door of a store.
   a. Front
   a. Rear
   b. Side
   c. No statement can be made regarding which door is easier to force, as this depends on the type of store and location.

9. When confronting a commercial building with a metal door that is set in a masonry wall and the door must be forced open, the quickest method would be
   a. Placing the adz end of a pry tool in the doorframe, drive it in, and force it outward
   b. Using a Rabbet tool
   c. Pulling the lock
   d. Removing exposed door hinges

10. A _____ lock is practically impossible to force.
    a. Rotating tumbler triple
    b. Straight double-bar bolt
    c. Detroit
    d. Fox

11. Sometimes it is necessary to breach a wall to enter an apartment. In this regard
    a. It is generally easier to breach the wall between the apartment and corridor.
    b. It is generally easier to breach the wall between apartments.
    c. Walls between apartments and corridors are of the same construction; thus, breaching either wall will require the same effort.
    d. Always breach the wall next to the door.

12. When a concrete block exterior wall is breached to gain entry
    a. Make the opening next to the door.
    b. Create an opening as near as possible to the center of the wall.
    c. Open the wall at a corner.
    d. Open the wall next to a corner.

d. Each of these presents special problems; therefore, none can be established as more or less difficult to force.

Aerial Operations

Chapter 8

Learning Objectives

- Identify an aerial fire apparatus as defined by NFPA 1901, *Standard for Automotive Fire Apparatus.*

- List the tasks that an aerial fire apparatus allows fire fighters to perform.

- Identify the stress factors that work against the strength of an aerial device.

- Review the proper guidelines to be used when rescuing a victim with an aerial fire apparatus.

- Review the proper guidelines to be used when performing ventilation techniques with an aerial fire apparatus.

- Review the proper guidelines to be used when operating hose lines from an aerial fire apparatus.

Aerial fire apparatus are resourceful yet complex pieces of equipment that allow fire fighters to operate at levels above and at times below grade. The operation of aerial fire apparatus does not come without hazards and dangers. Driver/operators must be competent not only in driving skills, but they must be proficient in the positioning, stabilization, and operation of the aerial device. Company officers, driver/operators, and crew members must be well trained in the manufacturer's recommendations relating to the proper operation of the aerial apparatus. All members of the company must be cognizant of the capabilities, as well as the limitations, of the aerial fire apparatus. They must understand what they can and cannot do.

Firefighting is a dangerous job with many harrowing situations that fire fighters may encounter at an incident. Working off or in an aerial device above ground can only add to the danger. Fire fighters must recognize the potential for injury when operating on the fire ground. The apparatus must be positioned properly on a solid surface for the intended operation, cautious of the fire building, as well as fire conditions. The apparatus must be stabilized according to manufacturer's recommendations to avoid tipping over. Care must be given when working around electrical wires, trees, or other overhead obstructions. Fire fighters must work safely while climbing and descending an aerial device, during rescue operations, when carrying equipment, or when performing other tasks. Weight limitations and other factors that cause stress to the aerial device must also be considered. Fire fighters must be involved in a comprehensive training curriculum, including a driver training program, to ensure that the driver/operator is proficient in the skill of driving and operating aerial fire apparatus.

Overview

Aerial fire apparatus is defined by NFPA 1901, *Standard for Automotive Fire Apparatus,* as a vehicle equipped with an aerial ladder, elevating platform, aerial ladder platform, or water tower that is designed and equipped to support firefighting and rescue operations by positioning personnel, handling materials, providing continuous egress, or discharging water at positions elevated from the ground.

Each type of aerial device—aerial ladder, elevating platform, or water tower—is designed with its own unique features. A fire department should determine which features would best provide for the particular work that is needed in its community or jurisdiction. Many factors may determine the final selection. Considerations should include the type of aerial fire apparatus, including the length of the aerial device, its required performance, the maximum number of fire fighters to ride within the apparatus, and additional equipment such as a fire pump, hose, or other provisions.

An **aerial ladder** is a self-supporting, turntable-mounted, power-operated ladder of two or more sections permanently attached to a self-propelled automotive fire apparatus and designed to provide a continuous egress route from an elevated position to the ground.

A **quint** is a type of fire apparatus with a permanently mounted fire pump, a water tank, a hose storage area, an aerial ladder, or elevating platform with a permanently mounted waterway, and a complement of ground ladders.

The primary purpose of this type of apparatus is to combat structural and associated fires and to support firefighting and rescue operations by positioning personnel-handling materials, providing continuous egress, or discharging water at positions elevated from the ground.

An **elevating platform** is a self-supporting, turntable-mounted device consisting of a personnel-carrying platform attached to the uppermost boom of a series of power-operated booms that articulate and/or telescope and that are sometimes arranged to provide the continuous egress capabilities of an aerial ladder. An elevating platform may consist of two or more folding boom sections whose extension and retraction modes are accomplished by adjusting the angle of the knuckle joints.

A **water tower** is an aerial device consisting of permanently mounted power-operated booms and a waterway designed to supply a large-capacity, mobile, elevated water stream. The booms can be of articulating design or telescoping design.

The working height of an aerial ladder is measured from the ground to the highest ladder rung with the ladder at its maximum elevation and extension. An aerial platform is measured from the ground to the top surface of the highest platform handrail with the device at its maximum elevation and extension. As noted previously, aerial fire apparatus are designed to allow fire fighters to perform the following tasks:

1. Provide access to upper floors. An aerial device will allow fire fighters to gain access to upper floors and roofs. They may be used to augment or replace interior stairways, elevators, or fire escapes. For access, they especially are useful in relieving the load on stairways, which may already be overcrowded with evacuating occupants or have been damaged by fire. If the fire is several stories up in the building, an elevating platform is ideal in transporting personnel to upper floors. They not only allow fire fighters to gain entry into a building, but also are used for egress from the structure.

2. Rescue. Fire fighters can gain entry to upper floors through windows and from porches, balconies, or the roof to begin a primary search, rescue victims, and remove them to safety.

3. Ventilation. Horizontal and vertical ventilation can be performed by fire fighters working off an aerial device, or they could be placed in a position, such as on a roof, to perform the operation.

4. Advancing/operating hose lines. An aerial device can be positioned to allow fire fighters to advance hose lines into a building for fire attack, as well as allowing them to operate hose lines from the aerial device itself. It also could be used to advance a supply line to an upper floor for use as a portable standpipe.

5. Elevated master streams. An aerial device should be capable of providing water through a master stream device on the fire ground. Where a prepiped waterway is provided, the waterway shall be capable of flowing 1,000 gpm. Where a prepiped waterway is not permanently installed, a ladder pipe with various sized tips and a bracket should be provided along with sufficient lengths of 3-inch hose, hose straps, and halyards to control the operation from the ground.

6. Transporting tools and equipment. Tools and equipment can be carried to upper floors via an aerial ladder or in the case of an elevating platform, placed in the platform and transported. This will save time, usually will require fewer personnel, and is less tiring for fire fighters performing the task.

Aerial fire apparatus also can be used for nonfirefighting tasks such as rescuing a victim from an elevated or below grade emergency situation. Rescue, ventilation, and hose operations are discussed in further detail in this chapter. Elevated master streams are discussed in Chapter 11.

Safety Considerations

On initial delivery of the fire apparatus, the contractor shall supply a qualified representative to demonstrate the apparatus and provide initial instruction to representatives of the purchaser regarding the operation, care, and maintenance of the apparatus and equipment supplied.

After delivery of the fire apparatus, the purchaser shall be responsible for the ongoing training of its personnel to proficiency regarding the proper and safe use of the apparatus and associated equipment. Drivers/operators must be well trained and proficient in the operation of the aerial fire apparatus. A fire department must maintain a comprehensive training program for all personnel who are responsible for the operation of the apparatus as defined in NFPA 1002, *Standard for Fire Apparatus Driver/Operator Professional Qualifications,* and NFPA 1500, *Standard on Fire Department Occupational Safety and Health Program.* NFPA 1451, *Standard for a Fire Vehicle Operations Training Program,* should be referenced in establishing such a training program.

Apparatus Positioning

The following information recommends the way in which an aerial fire apparatus should be spotted, or positioned, for various fireground operations. The apparatus movements and positioning discussed here are recommendations, not rules, and must be modified on the basis of the manufacturer's specifications concerning the maximum reach of the aerial device in relationship to the angle at which it is raised. In all cases, the operator must follow the practices that the manufacturer established of the particular aerial fire apparatus being used.

Stress on the aerial device can cause a disastrous outcome on the fire ground or other activity for which the aerial device is used. Stresses are factors that work against the strength of the aerial device and include the following:

1. Improper stabilization of the aerial apparatus

2. Operations in unparallel locations such as on uphill or downhill grades

3. Improper use of the aerial device, including erratic movement

4. An excessive degree of angle, either horizontally or vertically

5. Length of extension

6. Weight of the aerial device caused by personnel, victims, water, hose, or other equipment

7. Wind effects

8. Icing conditions

9. Exposure to heat

10. Contact or collision with an object, such as a building

11. Nozzle reaction from elevated streams

In addition, the condition of the fire building or exposure as well as the location and extent of the fire will determine the positioning of the truck. Positioning of the apparatus is also affected by the presence of electrical wires, trees, and any other overhead obstructions, as well as the type of ground surface and its condition. Weather conditions should also be taken into consideration. Command and company members will determine positioning based on these factors.

The apparatus must be completely stabilized before the aerial device is raised from its bedded position. Generally, this includes locking the brakes, chocking the wheels, and setting the stabilizers or outriggers. Proper stabilization of the aerial fire apparatus is critical. Stabilizers, along with the pads, must be deployed with every use of the aerial device. Stabilizers are used to prevent the aerial fire apparatus from tipping over when the aerial device is positioned away from the centerline of the chassis. Stabilizers, with the pads underneath, should be placed on solid, even terrain. They should be fully extended for proper operation and not "short jacked." With the stabilizers properly set, the aerial device shall be capable of being raised from the bedded position to maximum elevation and extension and rotated 90 degrees. Again, the manufacturer's recommendations must be followed, and all of the steps in the stabilizing procedure must be completed.

There will be times when aerial apparatus will need to be operated off an incline. The driver/operator must understand the manufacturer's recommendations for operation of the aerial fire apparatus when working on a grade. The operator should spot the turntable downhill from the point of operation to reduce stress on the aerial device. If the approach is from downhill, the truck should stop short of the building, and the aerial device should be raised over the cab. If the approach is from uphill, the truck should go past the building and the aerial device raised over the back of the apparatus.

The operator must be extremely careful when raising, extending, or rotating the aerial device near electrical wires. The aerial device should be watched constantly to see that it does not touch the wires **Figure 8-1**. If necessary, an officer or crew member should be positioned where both the aerial device and the wires, or other overhead obstruction,

can be seen in order to alert the operator of an impending problem.

An elevating platform with an articulating boom presents an additional problem. The knuckle of the boom must not make contact with wires or other overhead obstructions on the opposite side of the street from the fire building and from the platform **Figure 8-2**. The operator will most likely be watching the platform and might not see the position of the knuckle. This problem is most severe on a narrow street when the platform is not being raised very high and the reverse overhang is quite long.

Fire fighters who are not actually on the apparatus must be trained not to touch the apparatus while it is working. Anyone touching both the ground and the apparatus can be severely injured or killed by electric shock if the aerial device makes contact with a power line. Fire fighters in the platform or on an aerial ladder have been injured and killed when the aerial device touched an electrical wire.

Officers should be aware of areas in which electric wires may present a problem. This could be at most any location in the community or company's operating district. If necessary,

Figure 8-1 Aerial operators must use care when moving the aerial device near wires or other overhead obstructions.

Figure 8-2 Operators of elevating platforms with an articulating boom must watch for wires and other obstructions.

operational training sessions should be held in such areas to determine the safest and most effective ways to raise the aerial device, especially during the time on an emergency. This could be accomplished during preincident planning of buildings, such as target hazards.

The aerial device should not be overloaded with personnel or equipment. Here, again, the manufacturer's recommendations must be carefully followed. The rated capacity of the aerial device is the total amount of weight of all personnel and equipment that can be supported at the outermost rung of an aerial ladder or on the platform of an elevating platform with the waterway uncharged. In addition, officers and crew members must be aware of the rated capacity while discharging water through the full range of monitor or nozzle movements as permitted by the aerial manufacturer. It is important that the manufacturer clearly defines for the user the aerial device's rated capacity in various positions and operation modes. A load limit indicator or an instruction plate, visible at the operator's position, shows the recommended safe load at any condition of an aerial device's elevation and extension.

The driver/operator must know whether to operate the tip of the aerial ladder or the elevating platform in an unsupported, or cantilever position, or in a supported position. An unsupported position is one in which the tip of the aerial ladder or the elevating platform is not touching or resting on an object, such as a window sill. The supported position has the tip of the aerial ladder or the elevating platform resting on an object. The manufacturer's recommendations must be followed regarding the proper positioning of the aerial device. With the exception of the platform of an elevating platform, no one should be on the aerial device while it is being elevated, rotated, or retracted.

Rescue

Rescue of victims is the first priority on the fire ground. Search-and-rescue operations, if necessary, will begin as soon as fire companies arrive on the fire ground. Although it is best to remove occupants by way of interior stairways or fire escapes, if available, there are situations in which aerial fire apparatus must be used to remove people from a fire building. Occupants might be at the windows calling for help or appear ready to jump when ladder companies arrive. Conditions within the building might make it necessary for searching fire fighters to evacuate victims through upper-story windows, rather than through interior stairways.

Whatever the reason, the aerial device should be used for rescue when victims are within its reach. In such cases, the apparatus should be positioned for rescue on arrival. If possible, victims should be removed using a system of priorities. Those victims in most danger because of the present fire conditions should be removed first. Victims gathered in groups should follow. Any remaining victims in the fire area should then be rescued, followed by victims in exposed areas. Command must be made aware of changing conditions that may put victims in immediate danger as well as victims who may be panicked or threatening to jump.

Spotting the Turntable

In Chapter 2, positioning the aerial apparatus in relation to other fire apparatus was discussed in general terms. It must be remembered that aerial fire apparatus is equipped with an aerial device with a fixed maximum length. The goal is to get the turntable of the apparatus into a position that will allow the aerial device to be used most effectively. The ideal position of operation is when the aerial device is perpendicular to the objective.

When the victims are at a single window or at several windows that are close together, the turntable should be spotted close to the victims, as shown in **Figure 8-3a**. If the wind is blowing across the front of the building, the turntable should be located upwind from the victims. The smoke from the fire will then be carried away from the approach of the aerial device as it is raised toward the victims. Both the victims and the operator will be able to see the aerial device, and the victims will be removed into a smoke-free area. If the fire is upwind of the victims, the turntable must be spotted in the best possible position to get to the victims quickly.

The upwind turntable position is even more important when fire is issuing from windows below the victims, as shown in **Figure 8-3b**. With the turntable in this position, convected heat and embers will be blown away from the aerial device, the fire fighters, and victims using it. If a hose line is available, it should be directed at the fire in an attempt to keep it away from the aerial device. As a temporary measure, this will protect the aerial device, rescuers, and the victims. Using a hose line with a wide-angle fog pattern will push flames and heat away and have less chance of forcing someone off the aerial device. The use of a hose line also may be necessary when the fire is upwind of the victims.

When the victims are located at some distance from each other, the turntable should be spotted between them if possible, as shown in **Figure 8-3c**; that is, it should be approximately centered between the victims who are furthest from each other. This means that the aerial device may be operated in smoke part of the time, and the smoke will probably obscure the operator's view of some victims; however, it is still the best position under the circumstances. It would take too much time to spot the turntable upwind to remove some victims and then reposition

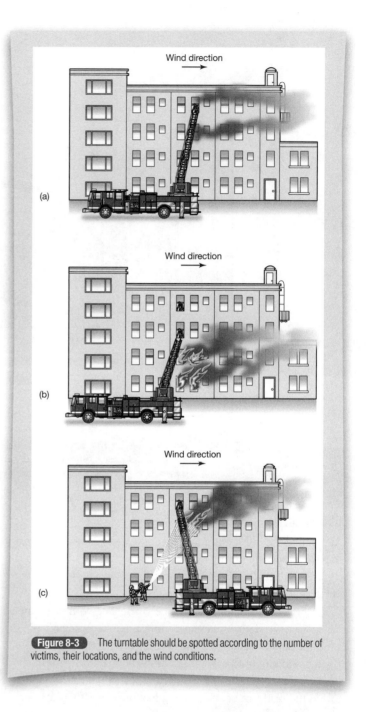

Figure 8-3 The turntable should be spotted according to the number of victims, their locations, and the wind conditions.

the apparatus to remove the rest. In addition, the arrival and positioning of other apparatus might make repositioning impossible.

Sometimes an aerial device can be positioned so that victims can be removed from two or more floors without repositioning the apparatus. In such situations, the turntable is spotted so that the aerial device can be raised parallel to the side of the building at an angle that will provide access to windows on more than one floor. In some cases, the aerial device is positioned at the corner of a building to permit coverage of two sides at the same time, as shown in **Figure 8-4**.

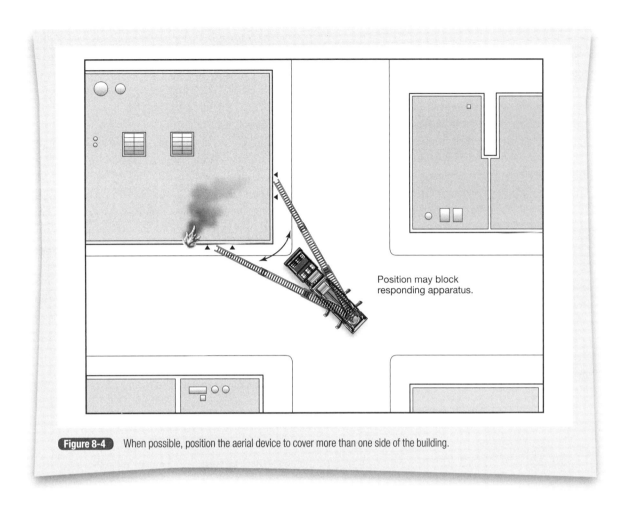

Figure 8-4 When possible, position the aerial device to cover more than one side of the building.

The company officer and apparatus driver/operator should work together to position the apparatus in the most advantageous location to perform rescue work. A quick sizeup of the situation on arrival or instructions from command should dictate how best to spot the aerial device. If the position of the turntable is not safe for aerial operations, including rescue, the apparatus should be repositioned.

Raising the Aerial Device

After the apparatus is spotted and stabilized, the aerial device should be raised toward the victims who are in the most danger. Normally, this means the people trapped closest to the fire; however, in some cases, occupants who are trapped at some distance from the fire might be in the most danger from the heat, smoke, and fire gases. They could be in the path of the products of combustion being carried by the wind, or these products may be flowing into their apartments or offices through interior shafts and stairways. In any case, the rescue situation must be sized up quickly and carefully and the victims in greatest danger removed first.

One other consideration affects the way in which the aerial device is raised. If the aerial device is extended toward occupants trapped at a particular window, then victims at higher story windows might attempt to jump to the aerial ladder or platform. This is especially likely when the people trapped above the extended aerial are panicky or when they are only one floor above the extended aerial device. Victims who jump for an aerial ladder or platform could kill or injure themselves, fire fighters, and other victims and could place the aerial device out of service.

An attempt should be made to establish visual and/or verbal contact with occupants who must await rescue while others are being removed. If they know that they are seen and will be rescued, they might calm down and stay in the building. Fire fighters must notify victims that they know they are in need of rescue and to stay where they are. Instructions should be clearly stated, and victims must be reassured to ease their fear. A bullhorn or other electronic device often is used to notify victims in need of rescue; however, smoke can obscure occupants' view of fire fighters, and the noise of the fire and firefighting operations might cover even the sound of a battery-powered bullhorn. Thus, the trapped occupants in the most peril are those endangered by their own mental state. Again, the situation should dictate which victims must be removed first.

In a normal approach to a window, fire escape, porch, or balcony, an aerial ladder is raised and rotated to the building.

The elevation is adjusted so that the tip of the ladder is aimed at or just above the sill or railing; then the aerial device is extended to the building. An elevating platform is normally extended so that it moves horizontally toward the window sill or balcony railing or up to it from below. These approaches should not be attempted during rescue operations because victims who are about to be removed from the building might attempt to jump down or across to the aerial device before it is in place.

When distance and height factors permit, the aerial device must first be raised well above the victims. Then the platform or aerial ladder tip should be lowered down to the victims in the final approach, as shown in **Figure 8-5**. In this way, the elevating platform or aerial ladder will not be approaching from below or be level with the victims until the last few seconds before they can reach it.

Placing the Aerial Ladder or Elevating Platform

During the final approach to the victims, the tip of the aerial ladder or the top rail of the platform should be placed carefully with respect to windows, fire escapes, porches, balconies, and roofs. The placement should allow trapped occupants to climb onto or into the aerial device with maximum ease and safety. It also should ease the job of fire fighters who are assisting or carrying victims onto the aerial ladder or into the elevating platform.

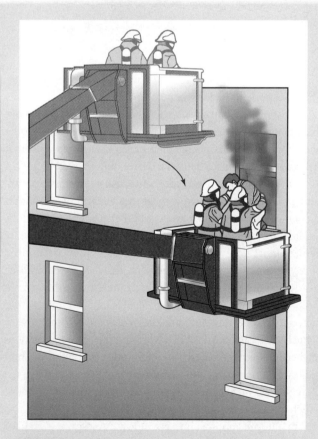

Figure 8-5   When used for rescue, the aerial device should be kept above the victims until the last seconds of the approach.

Aerial Ladders   If the aerial ladder tip is extended up and into a window, it will be in the way of victims and the fire fighters who are assisting them. Victims would have to climb onto the windowsill and swing over the top rail to get onto the ladder. Further, a ladder that is extended into a window might block too much of the window opening (as shown in **Figure 8-6a**) even if it is placed to one side of the opening. The tip of an aerial ladder should be placed so that the first rung is even with the windowsill. This allows victims to climb over the sill directly onto the ladder (as shown in **Figure 8-6b**) and allows fire fighters to pass a victim over the sill or railing quickly without having to maneuver the victim up and around the top rail.

When occupants are trapped on a fire escape or balcony directly over the fire, the situation is more acute. In such a case, the aerial ladder should be raised on the side of the fire escape or balcony that is least exposed to fire. In the final approach, the tip of the ladder should be placed 6 feet above the railing, as shown in **Figure 8-7a**. This will allow victims to use the top rail as handholds while they climb onto the ladder.

When removing victims from a flat roof, the tip of the aerial ladder should extend 6 feet above the edge of the roof, enabling victims to use the top rail as a handhold while climbing onto the ladder.

Elevating Platforms   The top rail of the platform should be placed even with the windowsill, as shown in **Figure 8-7b**. Again, this will allow the easiest and safest access to victims and to fire fighters assisting them. When a balcony or fire escape is above the fire, the platform should be raised on the least exposed side, but the top rail of the platform still should be placed about even with the railing. On flat roofs, the platform should be positioned so that it is just above and over the edge of the roof. This will allow direct access to the platform.

Imperfectly Spotted Turntable

It might be impossible to spot the turntable for a good final approach because of the objective being at an angle to the aerial device. Nevertheless, it should be spotted to allow as much as possible of the aerial ladder or elevating platform to make contact with the objective. This should also be attempted when the aerial device must be moved from one position to another.

Aerial Ladders   When only one beam of an aerial ladder will make contact with a window sill, fire escape, or balcony, the tip of the beam should be placed above and about 6 inches away from the objective during final approach.

This will allow the beam to settle, or rest, on the objective as victims climb onto the ladder. If the beam is placed on the objective, the weight of the victim or victims will cause the unsupported beam to twist downward. With both beams essentially unsupported, the fly section will bend evenly.

Wrong

Correct

(a)

(b)

Figure 8-6 When used for rescue, the tip of the aerial ladder should be placed so that the first rung is even with the windowsill.

Elevating Platforms When only one corner of a platform will contact the objective, the top rail of the platform should be placed a bit higher than the objective. Victims will then be able to get a good hold on the platform as they climb in. Fire fighters always should assist victims onto or into the aerial device used.

Removing Trapped Victims over Aerial Devices

When a number of occupants must be removed by an aerial ladder or elevating platform, they might try to force their way onto the aerial device. Fire fighters must guard against overloading, which can cause injury and death as well as failure of the aerial device. It would be beneficial to perform fire attack, ventilation, and rescue tasks, simultaneously in a coordinated fireground operation. These functions could improve conditions quickly. Fire fighters often can calm victims and get them to maintain their positions as conditions improve on the fire ground. If the situation does not improve quickly, previously calm victims can become panicky, and if rescues are to be made over aerial devices, then it will be all the more important that these rescue operations be performed quickly and effectively. One way to do so is to get additional ladder companies to the scene as soon as possible. Again, it is essential that the first-alarm response be increased in high life–hazard areas so that sufficient personnel and apparatus are available when needed.

Aerial Ladders At least one fire fighter should be assigned inside the building to assist occupants onto an aerial ladder being used for rescue. A fire fighter should be positioned with the victims, especially if they are grouped at one location and awaiting rescue. The fire fighter should direct victims to the ladder and then assist them onto the ladder in an orderly fashion. Fire fighters should guide able-bodied victims down the ladder, reassuring them as they move downward. Fire fighters should be cognizant of the ladder's rated capacity. Victims should be spaced out as they descend the aerial ladder. Manufacturer's recommendations must be followed to avoid overloading.

Small children and anyone unconscious or seriously injured must be carried down by fire fighters. For maximum safety, a backup fire fighter should precede the one carrying the victim— even if the victim is a child. The backup person can help the burdened fire fighter maintain balance, can hold the fire fighter against the ladder, can support some of the victim's weight, and if necessary, can direct the fire fighter in placing his or her feet on the rungs of the ladder, as shown in **Figure 8-8** .

When occupants must be removed from two or more locations, the aerial ladder should be kept at the first location until all the victims there have descended to the ground. Then the ladder can be moved to the next location. An aerial ladder should not be moved while anyone is climbing on it.

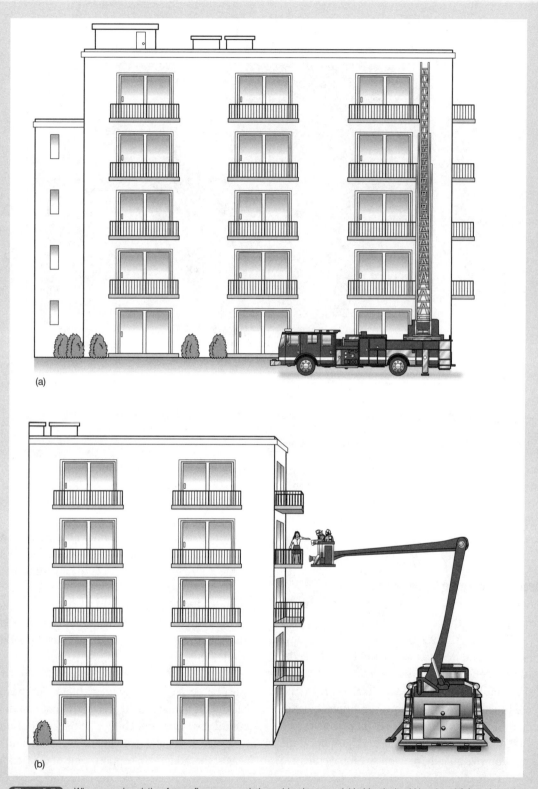

(a)

(b)

Figure 8-7   When removing victims from a fire escape or balcony (a) using an aerial ladder, it should be placed 6 feet above the railing, and (b) using an aerial platform, the top rail of the platform should be placed even with the railing.

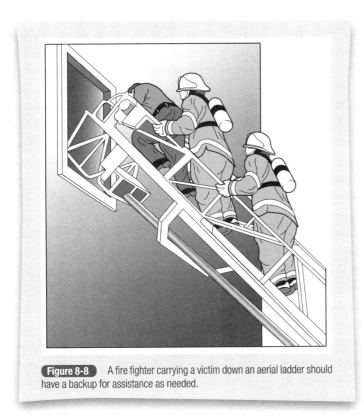

Figure 8-8 A fire fighter carrying a victim down an aerial ladder should have a backup for assistance as needed.

Elevating Platforms Removing victims in an elevating platform is easier and safer than over an aerial ladder, but the process tends to be slower because of frequent trips to the ground. At least one fire fighter should be assigned inside the building to assist victims and to make sure that the platform does not become overloaded. The situation can be difficult if the platform is occupied with victims who are being taken down while other victims await rescue. The fire fighter assigned to assist the victims must assure them of their safety and maintain order. As the platform returns, the fire fighter should control the loading area.

The removal of an unconscious victim requires a minimum of two fire fighters. One fire fighter is positioned to lift the victim to another fire fighter waiting in the platform. A fire fighter with an unconscious victim should protect the victim in the platform until it reaches the ground. Young children should be held low in the platform and protected, especially if they are panicky, until the descent is completed. Children and unconscious victims must be guarded when conscious adults are to be transported at the same time. Others could trample a child or an unconscious adult lying in the basket. If at all possible, unconscious victims should be sent down before conscious victims. They should be placed in the platform carefully and taken to the ground to receive medical attention. Fire fighters must assure that the platform is not overloaded whenever transporting victims, conscious or unconscious. Manufacturer's recommendations should be followed to avoid overloading.

Removing Victims by Litter

The situation may dictate that injured or unconscious victims be removed by litter. It is best to carry a litter down an interior stairway if the stairway is tenable and the victim is found on a lower floor; however, when the stairway is untenable and no other options are viable, the victim can be brought to the ground with an aerial device. A stokes basket or similar basket-type litter is preferred for such operations.

Aerial Ladders The easiest and safest method of removing a victim in a litter involves lowering the litter down the ladder between the rungs. One or two fire fighters are positioned on the ladder at the foot of the litter. Members will guide the litter down over the rungs. Additional fire fighters are positioned near the tip of the ladder and assist in the descent of the litter by controlling a rope attached to the head of the litter. Fire fighters positioned on the aerial ladder should be aware of the ladder's rated capacity.

Elevating Platforms The floor of the platform can be used to place a victim in a litter. If the injury requires that the victim cannot lie flat, the litter can be placed on a vertical plane and secured to the platform. If the litter will not fit in the platform, it could be laid across the top railing of the platform and secured with rope or appropriate strapping. During the descent, a fire fighter should maintain contact with the litter.

Aerial fire apparatus can be used in many ways to rescue victims from buildings. Efficient fireground operation will be promoted by preincident planning and inspections during which the turntable is spotted in different positions, and the movements that might be required are practiced.

Ventilation

An initial sizeup on arrival should determine whether the aerial device will be needed for ventilation or for another operations, such as rescue. In many fire situations, the ventilation of a multistory building begins at the roof. An aerial device can position fire fighters on the roof, but as noted in Chapter 5, other means could be used if they are available. If fire fighters are assigned to roof ventilation, the aerial ladder should be extended 6 feet over the edge of the roof. This will provide easy access onto the roof from the ladder and vice versa. In addition, it provides visibility to its location in darkness or smoky conditions.

If an elevating platform is being used, the bottom of the platform should be even with the roof or slightly over the edge. Fire fighters can exit and enter the platform easily from this position. Fire fighters should always check the integrity of the roof before stepping off the aerial device by sounding the surface with a tool. If a parapet wall is encountered, caution should be taken to keep the weight of the aerial device off the wall. The weight could cause the wall to collapse, especially if it is old and weakened by the elements. A roof ladder may be used for access from the top of the parapet wall to the roof deck below.

In addition to ventilation activities on the roof, the windows on the top floor or floors below may need to be opened or removed. The aerial device could be used for ventilation if the ladder crews are unable to reach upper floors by the interior stairways.

Removing Windows

An aerial ladder or an elevating platform can be used to position fire fighters to open or break windows for ventilation purposes. Fire fighters performing these tasks should be secured to the ladder or platform by a safety belt or harness to prevent any injuries while reaching out toward the window or by any sudden movements of the aerial device.

A leg lock should not be used when operating from an aerial ladder; if used, any unintentional extension or retraction of the fly section could cause serious injury. More then one fire fighter has been thrown off an aerial device in such a situation.

The use of a pike pole, plaster hook, or similar tool allows the fire fighter on the aerial device to operate several feet from the window to be opened. The fire fighter should be placed to one side of and slightly above the window, as shown in **Figure 8-9**. This will afford protection from heat, smoke, gases, and falling glass. If wind is a factor, placement should be on the windward side.

When more than one window must be opened and the wind is blowing across the face of the building, the first window opened should be the one furthest downwind if possible. Moving

Figure 8-9 Fire fighters can ventilate upper-story windows from aerial ladders or elevating platforms.

from window to window into the wind allows the combustion products leaving the windows to be carried away from the fire fighter, whose vision will not be obscured as it would be if upwind windows were opened first **Figure 8-10**.

Venting with Streams

There may be a situation where an elevated master stream could be used to break out the windows on upper floors. Solid streams should be used for this operation. The aerial device should be placed away from the building if a fire fighter is positioned on the aerial ladder or in the platform. This will protect the fire fighter from fire, combustion products, broken glass, and debris. As before, the furthest downwind window should be opened or broken first if the wind is blowing across the face of the building.

Venting with an Aerial Ladder

If it is known on arrival that the aerial device will be used for venting, the turntable can be spotted for maximum effectiveness. When the wind is blowing across the face of the building and exposures are located close to its downwind side, the turntable should be spotted just upwind of the closest exposure. This allows the aerial device to be used on the fire building and, if necessary, on the exposed building; however, if the fire building is relatively wide, the apparatus should be positioned closer to the center of the building.

If the fire building is very wide or if there are no downwind exposures, the turntable can be spotted at the center of the building to allow maximum reach and to put it within range of most top-floor windows. As noted earlier, aerial apparatus positioning must ultimately be determined by the existing conditions of the fire building or exposure, the location and extent of the fire, and the operation for which the aerial device is to be used.

As in other venting operations, the window furthest downwind should be opened first, and the aerial device should then be worked back into the wind. This will allow maximum visibility for the aerial operator and allow combustion products to be carried away from the aerial device; however, if fire issues through a window, the operator must move the aerial device away from it quickly. If the next lowest floor is also to be opened with personnel working from the aerial device, the top floor ventilation should be completed first. Then the aerial device should be moved to the furthest downwind window on the floor below and the procedure repeated.

Safety Precautions

Glass and debris will fall to the ground when windows are broken from an aerial device. Shards of glass and debris can slide down the aerial toward the operator, who must be aware of this possibility. In a strong wind, shards of glass can scale a good distance toward fire fighters performing other operations. Fire fighters should be warned before this operation, and command should ensure that fire fighters are kept clear of the immediate area.

Timing of the venting operation will be especially important if windows over the building entrances must be broken. Fire fighters, who most likely will be using the entrances to take hose

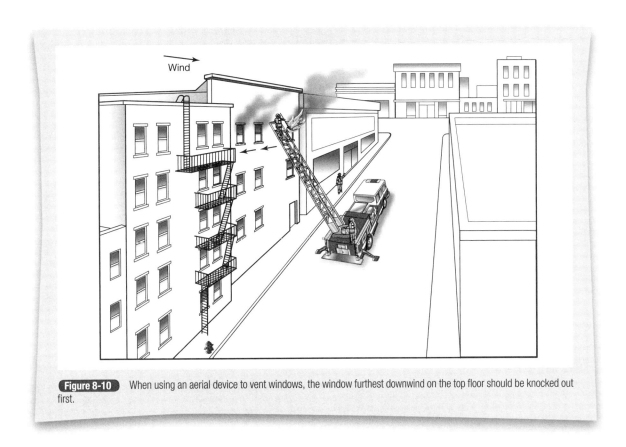

Figure 8-10 When using an aerial device to vent windows, the window furthest downwind on the top floor should be knocked out first.

lines and equipment into the building, should be warned to stay clear when the venting operation is to begin. Crews caught unexpectedly in a shower of glass and debris should keep their heads down with their arms close to their sides. They must not look up. They should move as close to the wall as possible, seek protection in doorways and under overhangs, and proceed only when advised that it is safe to do so **Figure 8-11**.

Hose Operations
Moving Equipment and Personnel
Aerial ladders and elevating platforms can be used to great advantage in getting personnel, hose lines, and equipment in the upper floors of a building. If an aerial ladder were used, fire fighters would need to carry this equipment up the ladder, or they could ascend the ladder and then hoist a hose line and equipment to their position. If an elevating platform were used, fire fighters would load the platform with the equipment for delivery to the upper floors or roof. Using an aerial device for this purpose usually is much faster than climbing stairs or ground ladders, especially if hose and/or equipment are needed above the third floor.

When fire fighters climb an aerial ladder carrying a hose line, the tip of the ladder should be placed even with the sill of the window they will enter. This placement is similar to that for rescue operations, but in this case, the ladder should be placed as close as possible to one side of the window to allow fire fighters, wearing full PPE and SCBA, to enter through the window.

If the ladder is placed so that two or three rungs extend over the sill, part of the window opening may be blocked. If the window opening is not large enough, they will have difficulty entering and their air cylinders could get hung up in the window opening. The entire window may need to be removed to provide sufficient access. Even with the aerial ladder properly placed, the window opening might not be large enough to allow easy access. If possible, fire fighters should choose a window that is large enough to afford easy access.

Supplying Water
Aerial devices can be used as a portable standpipe to supply water to the upper floors or to the roof. Modern aerial apparatus are usually equipped with a prepiped waterway system that eliminates the need of laying fire hose from the ground to upper levels over an aerial device. A prepiped waterway can be of a telescoping design. This system allows the waterway to extend from a swivel joint at the base of the aerial device to the tip of the last fly section. A nozzle is permanently fixed to the end of the aerial ladder or located in the platform of an elevating platform. Some elevating platforms have provisions for one or two 2½-inch discharge outlets. A hose line, or hose lines, can be taken off these discharge outlets and stretched into the building for use by firefighting crews. This system allows hose lines to be positioned on upper floors or to the roof of buildings, parking garages, bridges, or other elevated structures in a minimal amount of time.

Figure 8-11 Command should ensure that fire fighters are kept clear of the immediate area when window venting operations are being carried out.

Wrap-Up

Chief Concepts

- Aerial fire apparatus are versatile yet complex pieces of firefighting equipment. The aerial device can be used to perform several vital firefighting operations on upper floors and roofs of buildings, parking garages, bridges, and other overhead structures. An aerial device can be employed for search-and-rescue operations, ventilation and forcible entry operations, exposure protection, and the operation of elevated master streams.
- Driver/operators must be thoroughly trained in the operation of the equipment. They must be competent to drive, position, stabilize the aerial apparatus, and be proficient in the operation of the aerial device. In addition, they must be well versed in the maintenance and upkeep of the vehicle.
- Driver/operators, officers, and crew members should always operate the aerial equipment in accordance with the manufacturer's recommendations. They must know the capabilities and limitations of the apparatus and what they can and cannot do.
- Fire departments using aerial fire apparatus should be familiar with the following NFPA documents referencing this equipment:

  NFPA 1901, *Standard for Automotive Fire Apparatus*
  NFPA 1001, *Standard for Fire Fighting Professional Qualifications*
  NFPA 1002, *Standard for Fire Apparatus Driver/Operator Professional Qualifications*

  NFPA 1914, *Standard for Testing Fire Department Aerial Devices*
  NFPA 1451, *Standard for a Fire Service Vehicle Operations Training Program*
  NFPA 1500, *Standard on Fire Department Occupational Safety and Health Program*
  NFPA 1582, *Standard on Medical Requirements for Fire Fighters*

Key Terms

Aerial ladder: A power-operated ladder permanently mounted on a piece of apparatus.

Elevating platform: A self-supporting, turntable-mounted device consisting of a personnel-carrying platform attached to the uppermost boom of a series of power-operated booms that articulate and/or telescope, sometimes arranged to provide the continuous egress capabilities of an aerial ladder.

Quint: A type of fire apparatus with a permanently mounted fire pump, a water tank, a hose storage area, an aerial ladder, or elevating platform with a permanently mounted waterway and a complement of ground ladders.

Water tower: An aerial device consisting of permanently mounted power-operated booms and a waterway designed to supply a large capacity mobile elevated water stream.

1. When operating an aerial device on a downhill grade to rescue an occupant from an upper story window (target window), the aerial fire apparatus should be placed
   a. Even with the target window
   b. Uphill from the target window
   c. Downhill from the target window
   d. There is no recommendation for aerial fire apparatus placement when operating on a grade.

2. Fire fighters are in the greatest danger of electrocution when _____ an aerial fire apparatus that is in close proximity to wires.
   a. Working on the turntable of
   b. Working on the ladder of
   c. Inside the cab of
   d. Mounting or dismounting

3. No one should be on a(n) _____ while it is being rotated.
   a. Aerial ladder
   b. Anything
   c. Elevating platform
   d. Aerial ladder or water tower

4. When possible and conditions permit, it is best to remove victims from upper stories of buildings via
   a. Interior stairs and fire escapes
   b. Aerial ladders
   c. Elevating platforms
   d. Ground ladders

5. In most cases, it is best to position the aerial fire apparatus so the top of the ladder is upwind from the products of combustion; however, there may be exceptions when
   a. Victims are located at more than one window on a floor where aerial rescues are necessary.
   b. Victims are located on more than one floor.
   c. Victims are located on more than one side of the building.
   d. There are no exceptions to the upwind rule.

6. When placing an aerial ladder to rescue victims from a window, the top rung of the ladder should be placed
   a. Inside the window
   b. Even with the windowsill
   c. Alongside and above the windowsill on the windward side
   d. Alongside and above the windowsill on the leeward side

7. When placing an aerial ladder to rescue victims from a balcony, the top rung of the ladder should be placed
   a. To the front of the balcony, level with the balcony rail
   b. To the side furthest from the fire and even with the balcony rail
   c. To the front of the balcony and extended 6 feet above the balcony rail
   d. To the side that is least exposed to fire and extended 6 feet above the balcony rail

8. When placing an elevating platform to rescue victims from a balcony, the top rail of the platform should be placed
   a. To the front of the balcony, level with the balcony rail
   b. To the side that is least exposed to the fire and even with the balcony rail
   c. To the front of the balcony and 2 feet below the balcony rail
   d. To the side furthest from the fire and extended 2 feet below the balcony rail

9. An advantage of an aerial ladder over an elevating platform when removing many occupants from the same area is
   a. That it is safer
   b. That it is easier
   c. That it is quicker
   d. Actually, the elevating platform is safer, easier, and quicker than using an aerial under all circumstances.

10. When ventilating windows from an aerial ladder, the fire fighter should be secured to the ladder. Which of the methods listed below is an **unacceptable** means of securing the fire fighter to the ladder?
    a. Safety belt
    b. Harness
    c. Leg lock
    d. All of the above are acceptable ways to secure the fire fighter to a ladder.

11. When several windows on the same floor must be vented from an elevating platform, start with the window that is
    a. Nearest the elevating platform
    b. Farthest from the elevating platform, but still within reach of the pike pole being used
    c. Furthest upwind if possible
    d. Furthest downwind if possible

12. When fire fighters climb an aerial ladder carrying a hose line, the tip of the ladder should be placed _____ the window they will enter.
    a. Even with the sill and directly below
    b. Even with the sill and to one side of
    c. Two full rungs above and to the side of
    d. With two full rungs inside

Ground Ladders

Chapter 9

Learning Objectives

- Identify the several types of ground ladders.

- Define tasks that may be accomplished while using ground ladders.

- Identify the appropriate and safe procedures for carrying ground ladders.

- Identify the appropriate and safe procedures for raising ground ladders.

- Identify the appropriate and safe procedures for climbing ground ladders.

- Review the proper guidelines to be used when rescuing a victim over a ground ladder.

- Review the proper guidelines to be used when performing ventilation and forcible entry from a ground ladder.

- Review the proper guidelines to be used when advancing and operating hose lines from a ground ladder.

Ground ladders are used at an incident for rescue, ventilation, forcible entry, advancing hose lines, and fire attack. To accomplish these operations safely, fire fighters must be well trained and capable of carrying, raising, and climbing ground ladders. Fire fighters must have a basic knowledge of ground ladder construction and should be able to select the appropriate ladder for each job. They need to practice correct lifting and raising techniques and not overexert themselves. Fire fighters should use leg muscles but not back and arm muscles when lifting. There should be an adequate number of personnel available to carry and raise a ground ladder.

Ground ladders must be placed in locations that provide the optimum safety for fire fighters performing a specific task. When raising a ladder to the vertical position, a check of the overhead should always be made to check for obstructions such as trees or building overhangs and electrical power lines. Ground ladders may be made of wood, fiberglass, or aluminum. Most fire departments now use aluminum ladders. Ground ladders are made of aluminum, fiberglass, and wood. Ladders will conduct electricity and electrical power lines should always be avoided. Butts of ladders should be on a firm level surface and snow, ice, and other slippery surfaces should be avoided. After a ground ladder is in position, it should be secured to the building to prevent slippage. The ladder can be secured at the tip by using a rope hose tool, safety strap, or dog chain. The base of the ladder can also be secured by attaching the base of the ladder to a fixed object. It is important that fire fighters do not forget the basics. They must always foot ladders when raising them and they should do this as a team. Depending on the height of the ladder, this will determine how many personnel are required to complete a safe lift.

For the safety of the fire fighter and to provide the optimum combination of load carrying and stability, ground ladders must be set at the correct angle of inclination. A simple formula for the correct placement is to position the bottom of the ladder at a distance from the vertical plane equal to one fourth the total working length of the ladder. Another simple and more practical test is for the fire fighter to extend their arms after they stand on the ground or first rung of the ladder. Their arms should reach a rung parallel to their body without stretching or straining. If the distance exceeds their reach, more than likely, the ladder position needs to be adjusted.

Before climbing an extension ladder, ladder locks or dogs must be properly seated and halyards need to be tied off to the ladder base. A fire fighter should heel the ladder by placing a foot on the beam at ground level or by standing behind the ladder and pulling backward toward the building. Fire fighters should use proper climbing techniques, climb the ladder smoothly and rhythmically, and maintain contact with the ladder at all times. The ladder should be positioned so that contact is maintained with the ladder when dismounting. Always ensure that the area is stable and free of tools and equipment and that it is safe before stepping away from the ladder.

Ground ladders should not be overloaded. Only one person should be allowed on each section of a ground ladder at the same time. NFPA 1932, *Standard on Use, Maintenance and Service Testing of Fire Department Ground Ladders,* establishes ladder loading requirements. Removing occupants out of a building over ground ladders is not an easy task. Although this operation may need to be accomplished, fire fighters should first determine whether there are alternate methods, such as interior stairways or the availability of elevating platforms, to remove the occupants.

When working on a ladder, fire fighters should pay particular attention to their immediate environment. This includes areas above and the below where they are positioned. A fire fighter should be secured to the ladder by a fall protection harness, safety belt, or a leg lock when they are working or in a stable position and not climbing the ladder.

A ladder may be damaged at an incident because of direct contact with the fire, or it may have been dropped, overloaded, or otherwise destroyed. Ladders that have been damaged should be placed out of service and should not be used. Ground ladders need to be inspected, cleaned, maintained, and service tested on a regular and set timetable by the department so that they will function as designed and not compromise the safety of fire fighters during an emergency.

Ground ladders have been the mainstay of fire departments across the country for years. Even with aerial ladders and platforms readily available, ground ladders maintain an integral role in ladder company operations. Many fire departments use only ground ladders, especially these departments located where buildings are not more than three or four stories in height.

The main advantage of a ground ladder is its portability. It can be carried to positions that may not be reached with aerial apparatus. Where a ladder company has both ground ladders and an aerial device, the ground ladders can be used for lower-story operations, freeing the aerial device for upper-story work or any distance in excess of the reach of ground ladders. As a rule, anywhere you have buildings in your response area that exceed the height of your ground ladders, you have high-rise buildings!

Aerial apparatus carry a sufficient number of ground ladders in various lengths that can be raised at different locations simultaneously. In contrast, an aerial device must be raised from a fixed position once the truck has been set up. The aerial device may have limited reach from this position for rescue or fire suppression.

The major disadvantages of ground ladders are their limited reach and the personnel required to carry, raise, and climb them. Ground ladder operations start to become inefficient and time consuming when the ladders must be long enough to reach above three floors. Depending on its length, from two to six members may be needed to raise the ladder as compared with the single fire fighter required to raise an aerial device. Staffing levels may prohibit the raising of ground ladders in a timely fashion.

Types of Ground Ladders

Construction Features

Several different types of ground ladders exist, each having a specific function. Three primary materials are used in the construction of fire department ground ladders. The aluminum alloy ladder has become the most popular ladder used in the fire service. This is due to its strength-to-weight characteristics, durability, and cost. Wooden ladders as well as fiberglass ladders are manufactured and also are used in the fire service.

Basically two types of construction methods are used for ground ladders. Truss construction has an open side rail or beam while the solid-beam construction does not. NFPA standards do not differentiate between the types of construction when defining strength requirement and choosing one type of construction from the other is usually based on a fire department's needs or past practices.

Classification Features

Fire department ground ladders can be classified as follows:
- Single or wall ladders
- Roof ladders
- Extension ladders
- Folding ladders
- Combination ladders
- Pompier ladders

Single or Wall Ladders   A single, or wall, ladder is designed in one section only. It is nonadjustable in height and generally come in heights of 10, 12, 14, 16, 18, 20, and 24 feet.

Roof Ladders   **Roof ladders** also are designed in one section only. They are equipped with folding hooks at the tip to provide an anchoring mechanism when the ladder is placed over the peak or ridgepole of a building. They are designed to lie flat on a roof

as a work platform. Roof ladders can also function as a single ladder if required. They generally come in heights of 8, 10, 12, 14, 16, 18, and 20 feet.

Extension Ladders   An **extension ladder** is a ladder that is adjustable in height. It has a base ladder and one or more fly sections that travel in guides allowing for extension of the fly section. The two-section ladder consists of the base ladder and one telescoping fly section, whereas the three-section ladder consists of the base ladder and two telescoping fly sections. Two-section extension ladders generally come in heights from 20 to 50 feet, whereas three-section extension ladders range from 24 to 50 feet, depending on the construction of the ladder. Longer ladders, those from 40 to 50 feet, are equipped with staypoles.

Folding Ladders   A **folding ladder,** or attic ladder, is a single ladder that has hinged rungs. This allows the ladder to be "folded" so that the beams rest against each other. Folding ladders are used in tight spaces such as narrow corridors or walkways and can be used in small spaces such as closets or attic scuttle holes. Folding ladders generally come in heights of 8, 10, 12, and 14 feet.

Combination Ladders   A **combination ladder,** or A-frame ladder, has a variety of applications. It can be used as a single ladder, an extension ladder, and as an A-frame, which is similar to a stepladder. These ladders generally come in heights of 10, 12, 14, and 16 feet.

Pompier Ladders   The **pompier ladder** consists of a single-beam ladder with rungs protruding from either side. There is a large hook at the top that is placed on the sill in an open window or other opening. The ladder is used to climb from floor to floor by way of the window openings. This ladder is not the easiest means of gaining access to upper floors and may not be the safest manner to do so. This ladder is usually found in 16-foot lengths. Pompier ladders are used rarely by most fire departments today. They are used in some cases for training to build confidence and basic training.

Ground Ladder Operations

Within their height and personnel limitations, ground ladders can be effective in a number of firefighting operations. The following tasks may be accomplished using ground ladders:

- Gain access to upper floors of the fire building and exposure buildings
- Remove victims trapped on upper floors
- Advance hose lines to upper floors when stairways are being used by occupants escaping the building
- Advance hose lines to upper floors when fire fighters are using stairways for other fire department operations
- Replace damaged stairways to upper floors
- Remove occupants from crowded fire escapes
- Obtain access from one roof level to another
- Bridge fences, narrow walkways, courts, and alleys
- Provide proper ventilation techniques
- Transport an injured victim if proper equipment is not available
- Use for assistance in salvage and overhaul operations

The utilization of ground ladders is discussed, by operation, in the following sections. The remainder of this chapter discusses ground ladder operations in general. Ladder company members are usually assigned to tasks on the fire ground that involve the use of ground ladders. Aerial apparatus will carry an assortment of ground ladders for use by fire department personnel; however, many departments do not possess aerial apparatus. Ground ladders will be carried on pumpers, tankers, utility apparatus, or other designated vehicles. Departments without assigned ladder company personnel should have a standard operating guideline, which mandates a procedure to be initiated when ground ladders need to be deployed on the fire ground or at other incidents.

If laddering duties are not specifically assigned, rescue operations or fire attack operations can be unnecessarily delayed. For example, if too few fire fighters are assigned to raise ladders to upper floors, ventilation, forcible entry, or the primary search and rescue operations could be delayed. Crews prepared to make entry into the building will need to wait for ladders to be raised before these tasks can begin. If too many fire fighters decide to raise ladders, too few attack lines will be advanced to support the primary search and extinguish the fire.

Thus, whether or not a department consists of separate engine and ladder companies, the incident commander must ensure that fire fighters are assigned specific tasks on the fire ground and ensure that ground ladder operations are being performed as needed. Engine and ladder company personnel should hold combined training sessions because they must support a coordinated fire attack on the fire ground. In particular, ladder operations can affect the efficiency of other fireground operations.

Ground ladders are available in a number of different construction types and lengths. NFPA 1901, *Standard on Automotive Fire Apparatus,* provides a minimum requirement for the types of ladders required on different classifications of fire vehicles, including aerial apparatus and pumpers. Ground ladders carried on fire apparatus will depend on standard recommendations, the types of structures that comprise the response area of the community, and the department's past experience.

Carrying Ladders

There are several ways to carry a ground ladder from the apparatus to a building. The method used in a particular instance should be the one that requires the least maneuvering and the least amount of time. Depending on the type of fire apparatus, ground ladders may be mounted in a number of different configurations and locations. They may be located on the sides of pumpers, stored above reach and lowered mechanically, or nested on runners and removed from the rear of aerial apparatus. No matter how ground ladders are stored, fire fighters must know how to remove them expeditiously and place them in service.

Ladders can be carried in a variety of ways depending on the length of the ladder and the proper number of fire fighters required to perform the task. Short, lightweight ladders may be carried by one fire fighter using the low shoulder carry or the high shoulder carry. Two fire fighters may also carry a ground ladder using the low shoulder method from flat or vertical mounting or when picked up from the ground. Other techniques that may be used by two fire fighters are the underarm and arm's length carry.

Key Points

Departments without assigned ladder company personnel should have a standard operating guideline, which mandates a procedure to be initiated when ground ladders need to be deployed on the fire ground or other incident.

Key Points

There are several ways to carry a ground ladder from the apparatus to a building. The method used in a particular instance should be the one that requires the least maneuvering and the least amount of time.

Key Points

Ladders can be carried in a variety of ways depending on the length of the ladder and the proper number of fire fighters required to perform the task.

As ground ladders get longer and heavier, additional personnel will be needed to carry them. Three fire fighters could use the flat shoulder carry from flat mounting positions or when picked up from the ground. Another variation is the flat arms length carry. The flat shoulder carry can be used with slight variations in position for four and six fire fighter carrying methods. When encountering a narrow opening such as the area between parked automobiles or a narrow walkway, an overhead carry will enable fire fighters to proceed through the opening.

Figure 9-1 shows some of the ways in which ladders can be carried depending on the number of fire fighters and the length of the ladder.

Raising Ground Ladders

Ground ladders carried to the fire ground must be properly positioned before they are raised. Ground ladders may be used for rescue, ventilation, forcible entry, or fire attack. It is advantageous to place the ladder and raise it properly for its intended use. In

Figure 9-1   Ladders carried in flat position are easily handled and quickly raised.

this way, the ladder will not have to be relocated or repositioned for a better climbing angle.

Raising ladders can be accomplished using one fire fighter for straight-wall ladders and short-extension ladders to several fire fighters for larger extension ladders. A single fire fighter should be able to raise a straight-wall ladder by placing the butt against the ground at a proper distance and then walking the ladder to a vertical position before placing the tip against the building. The ladder butt could also be placed up against the base of the building, walked to a vertical position, and then the butt pulled from the building to acquire the proper climbing angle. Depending on the strength of the fire fighter, a short-extension ladder may be raised from the high shoulder carry and the low shoulder carry methods.

It is much easier to raise an extension ladder using two fire fighters to perform the task. Two methods of raising the ladder using two fire fighters are the flat raise and the beam raise. In the flat raise, one fire fighter, positioned at the butt of the ladder, places the ladder on the ground while the other fire fighter maintains a position near the tip. The fire fighter at the butt heels, or anchors, the ladder in position by standing on the bottom rung. The heeler then crouches down and grabs a rung or the beams and leans back while the fire fighter at the tip steps beneath the ladder and raises it hand over hand to the vertical position. In this position, both fire fighters place their toes against the same beam to steady the ladder.

If the ladder needs to be pivoted, or turned, so that the fly section is to the outside, it should be done before the ladder

is extended. If the ladder is to be extended, the fire fighter on the inside of the ladder grasps the halyard and extends the fly section to the proper height, making sure that the ladder locks are secure. The ladder is then slowly lowered onto the building with the fire fighter on the outside resting his foot against the butt spur or the bottom rung.

In the beam raise, the fire fighter at the butt end of the ladder places one beam on the ground while the other fire fighter at the tip rests the beam on his or her shoulder. The fire fighter at the butt grasps the beam of the ladder with both hands and leans backward, acting as a counterweight. The fire fighter at the tip grasps the beam and raises the ladder hand over hand until in the vertical position. After the ladder is in the vertical position, the same sequence used in the flat raise is followed to lower the ladder onto the building.

A minimum of three fire fighters is usually required to raise extension ladders 35 feet or larger. As with two fire fighter raises, both the flat raise and the beam raise can be used. The raising of an extension ladder using the flat raise is very similar to the two fire fighter raise with the addition of the third fire fighter at the tip end. Both fire fighters at the tip advance hand over hand in unison until the ladder is in the vertical position. They both steady the ladder while the third fire fighter grasps the halyard and extends the fly section. The beam raise is similar to the two fire fighter raise. The only difference is that the third fire fighter is also positioned along the beam as the ladder is being raised.

If additional personnel are available, a fourth member could be used to help raise the heavier extension ladders. Using the flat raise method, two fire fighters heel the ladder after the butt has been placed on the ground and two fire fighters at the tip advance in unison until the ladder is in the vertical position. The ladder can be pivoted, raised, and lowered onto the building with a coordinated effort of all four fire fighters.

Most ground ladder raises can be accomplished without placing the ladder on the ground initially. An exception would be ground ladder of 40 feet or more with staypoles attached to each beam to provide stability when the ladder is raised. Fire

Key Points

Fire fighters should remove the ladders from the apparatus and carry them with the butt end toward the building.

fighters should remove the ladders from the apparatus and carry them with the butt end toward the building.

After the butt is in position near the building wall, it is lowered to the ground while the tip of the ladder is raised from shoulder height. The ladder then can be pivoted and extended if necessary. With practice, this raise can be performed in one smooth nonstop motion.

Two- and three-person ladder raises also can be quickly applied using the beam raise. The ladder is turned with its rungs parallel to the building wall as it is lowered to the ground. The butt of one beam is footed, and the ladder is raised to the vertical position. This raise is often advantageous when wires, trees, or other obstructions restrict the area above the ladder position.

Extension ladders should be used in the fly up, fly away from the building position unless otherwise specified by the manufacturer. This places the halyard between the ladder and the building, and thus, the ladder will move toward the building as the halyard is pulled to extend the fly section, as shown in **Figure 9-2**. If the ladder should get away from the fire fighters, it should fall against the building and not out onto crews and apparatus.

Angle of Inclination

In order to provide the optimum combination of load carrying and stability, ground ladders must be set at the correct angle of inclination by positioning the base section a horizontal distance from the vertical wall equal to one quarter the working length of the ground ladder, according to NFPA 1932, *Standard on Use, Maintenance, and Service Testing of In-Service Fire Department Ground Ladders*. An angle of inclination of between 70 degrees and 76 degrees is permitted, with an angle of 75.5 degrees being optimum. Extreme caution should be used when the angle of inclination is less than 70 degrees.

A simple formula for the correct placement is to position the bottom of the ladder at a distance from the vertical plane

Figure 9-2   Maximum safety is obtained by extending a ladder with the fly section on the outside.

equal to one quarter the total working length of the ladder. At this angle, which is shown in **Figure 9-3**, the ladder will give the maximum strength.

Safety Considerations

Fire department ground ladders should be used for rescue, firefighting operations, and training and should not be used for any other purpose. Ground ladder butts should be set on a firm, level base before ground ladders are used. Ground ladders should not be placed on ice, snow, or slippery surfaces unless means to prevent slippage are employed.

Extreme caution should be used when work occurs around charged electrical circuits because metal ground ladders conduct electricity. All metal ground ladders should be kept away from

Key Points

Extension ladders should be used in the fly up, fly away from the building position unless otherwise specified by the manufacturer. This places the halyard between the ladder and the building, so the ladder will move toward the building as the halyard is pulled to extend the fly section.

Key Points

Ground ladder butts should be set on a firm, level base before ground ladders are used. Ground ladders should not be placed on ice, snow, or slippery surfaces unless means to prevent slippage are employed.

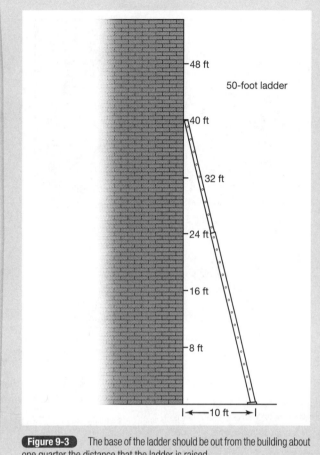

48 ft

50-foot ladder

40 ft

32 ft

24 ft

16 ft

8 ft

|← 10 ft →|

Figure 9-3   The base of the ladder should be out from the building about one quarter the distance that the ladder is raised.

power lines or other potential electrical hazards. Wood and fiberglass ground ladders also can conduct electricity, and extreme caution should be used around electrical hazards.

Ladder crews raising ground ladders should be careful of overhead obstructions, especially electrical wires. A check of the overhead always should be conducted before raising a ladder to the vertical position. Fire fighters should watch the tip of the ladder, not the butt, as they are raising it. A very small movement of the butt can result in a wide swing at the tip; the longer the ladder, the more pronounced this effect is. Ladder locks must be seated before the first person climbs the ladder.

Ground ladders should be secured at the top to prevent slippage by the first person to climb the ladder. After the ladder is raised, it should be tied in to the building. Ground ladders should

Key Points

A check of the overhead should always be conducted before raising a ladder to the vertical position.

Key Points

Ground ladders should be secured at the top to prevent slippage by the first person to climb the ladder.

Key Points

Ground ladders should be secured at the base either by a fire fighter or by mechanical means to prevent slippage.

be secured at the base, either by a fire fighter or by mechanical means, to prevent slippage. Either action will keep the ladder from moving while being used. Halyards on extension ladders should be tied off to the base section or should be otherwise secured before the ground ladder is climbed.

A ground ladder should not be overloaded. NFPA 1932, *Standard on Use Maintenance, and Service Testing of In-Service Fire Department Ground Ladders,* establishes ladder loading. The total weight on the ground ladder, including people, their equipment, and any other weight, such as a charged fire hose, should not exceed the duty rating as given. Single and roof ladders, all extension ladders, and combination ladders have a maximum load of 750 pounds. Folding and pompier ladders have a maximum load of 300 pounds. These weights only are applicable for ladders set at a proper angle of inclination for climbing.

After a ladder has been used to enter a building, it should be left in place as an exit, unless it is required for rescue at a nearby location. It then should be returned to its original position as soon as the rescue has been effected. Fire fighters should be made aware if the ladder has been removed, as this may be their only means of escape from the building.

Key Points

The total weight on the ground ladder, including people, their equipment, and any other weight, such as a charged fire hose, should not exceed the duty rating as given.

Key Points

After a ladder has been used to enter a building, it should be left in place as an exit, unless it is required for rescue at a nearby location. It should then be returned to its original position as soon as the rescue has been effected.

Key Points

Ladders can be placed at windows, fire escapes, or balconies and roofs so that occupants can climb or be carried down.

Key Points

No matter how ground ladders are to be used for rescue, the first ladders should be raised to the victims in most danger.

Rescue

Ground ladders are used in several ways to help remove people from a building. The ladders can be placed at windows, fire escapes, balconies, and roofs so that occupants can climb or be carried down.

No matter how ground ladders are to be used for rescue, the first ladders should be raised to the victims in most danger. As previously noted, these are usually the victims closest to the fire on or directly above the fire floor; however, smoke and other combustion products may be endangering victims further away from the fire so that they, too, may require immediate attention. Observation of the fire situation will indicate where the first ground ladders should be raised.

Ladder Placement

It is important that the ladder crew selects a ladder of the proper length for the raise—especially when a single ladder, rather than an extension ladder, is being used. There is little sense in tying up a longer ladder for a comparatively short raise. This could strand victims on upper floors who cannot be reached with the remaining shorter single or extension ladders. In addition, a ladder that is too long for the raise must be placed at an awkward climbing angle or placed so that it blocks part of the window opening.

Training sessions at buildings in the response area will help in determining which ladders might be needed at different locations. This depends on the building's floor heights, which can vary from 8 to 20 feet or more depending on construction features.

Ground ladders can be placed for rescue, ventilation, forcible entry, and fire attack. As a ground ladder is raised for rescue, it

Key Points

It is important that the ladder crew selects a ladder of the proper length for the raise—especially when a single ladder, rather than an extension ladder, is being used.

Key Points

As a ground ladder is raised for rescue, it must be initially kept out of the reach of people who are to be removed. This is especially important when the ladder is being raised past victims to reach others on the floor above.

must be initially kept out of the reach of people who are to be removed. This is especially important when the ladder is being raised past victims to reach others on the floor above, as shown in **Figure 9-4**. It is almost impossible to place a ladder properly while someone has hold of the upper part or tip. The ladder should be raised in a vertical position away from the building, pivoted, extended if necessary, the tip lowered to the victims,

Figure 9-4   When raising a ground ladder for rescue, the ladder should be kept out from the building so that victims cannot get hold of the ladder and hinder operations.

Key Points

When a ground ladder is placed at a window for rescue, the size of the window opening must be taken into consideration. The tip should usually be at or just over the level of the sill.

and the butt quickly moved to the proper position. Fire fighters must ensure victims in peril as to their intentions for rescue with commands that will be understood and followed.

When a ground ladder is placed at a window for rescue, the size of the window opening must be taken into consideration. The tip should usually be at or just over the level of the sill, as shown in **Figure 9-5**. If the window opening is large enough, the ladder should be placed inside the window two or three

Figure 9-5   For rescue, ground ladders should be placed with the top rung at or just over the sill.

Key Points

In placing a ground ladder at the front railing of a balcony or fire escape or at the wall beside either of these features, two to four rungs should extend above the railing. This will provide a good handhold for victims climbing onto the ladder and for fire fighters working with victims.

rungs if possible and secured before effecting the rescue. A large window opening will also allow fire fighters to enter and exit the building with less effort.

In placing a ground ladder at the front railing of a balcony or fire escape or at the wall beside either of these features, two to four rungs should extend above the railing. This will provide a good handhold for victims climbing onto the ladder and for fire fighters working with victims. This gives victims a greater sense of security, as do the trusses on an aerial ladder. Ladders raised to a roof should also extend four to five rungs above the roof wall for the same reason.

Ground Ladders as Exits

If ground ladders will reach to floors being searched for victims, they should be raised as a means of egress for fire fighters with victims and as emergency exits. This should be done even if fire fighters engaged in the primary search have entered the building through interior stairways. The ladders provide additional exits in areas somewhat removed from the stairways and allow fire fighters to get victims to fresh air quickly without having to work their way through the building. The ladders also provide fire fighters and victims with alternative exits in case the stairways become impassable.

Ground ladders used for egress should be raised at corridor windows if their locations are known. Alternate

Key Points

If ground ladders will reach to floors being searched for victims, they should be raised as a means of egress for fire fighters with victims and as emergency exits. This should be done even if fire fighters engaged in the primary search have entered the building through interior stairways.

Key Points

Ground ladders used for egress should be raised at corridor windows if their locations are known. Alternate positions include balconies, porches, other windows, and roofs. Command should notify fire fighters as to the placement of these ladders.

positions include balconies, porches, other windows, and roofs. Command should notify fire fighters as to the placement of these ladders.

Radio communications are helpful in these operations. Fire fighters inside can use their portable radios to call for ladders at certain locations. The officer in charge can use the radio to advise those inside that ladders have been placed at certain positions. Where radios are not available, visual and verbal contact, initiated by fire fighters working inside, can be used to properly locate ladders. As noted earlier, after a ladder has been placed and fire fighters are aware of its location, it should not be removed without authorization and notification of its removal.

Rescuing Victims from a Ladder

Rescuing a victim over a ground ladder is by no means an easy task. A conscious victim may be in a near panic state, and the thought of climbing down a ground ladder from any height above the ground may add to the anxiety. Conscious victims must be constantly reassured by the fire fighters that they will be safely removed from the peril.

If possible, at least one fire fighter should be inside the building to assist the victim onto the ladder. A fire fighter on the ladder supports and encourages the victim as they descend the ladder. The fire fighter should place his or her hands under the victim's armpits and maintains contact with the rungs of the ladder. If the victim slips, the fire fighter should be able to maintain control. Children should be carried by the fire fighter and not allowed to climb down the ladder because of safety considerations.

Removing an unconscious victim over a ground ladder is a very difficult task. Complicating the rescue is the size and weight of the victim compared with that of the fire fighter. Before this rescue is attempted, alternative ways of removing the victim from the building should be considered. Interior stairways are the most reliable method of removal and should always be used if available. Aerial ladders, fire escapes, and adjacent areas in the building or roofs are alternative methods.

> **Key Points**
>
> If possible, at least one fire fighter should be inside the building to assist the victim onto the ladder. A fire fighter on the ladder supports and encourages the victim as they descend the ladder.

> **Key Points**
>
> Removing an unconscious victim over a ground ladder is a very difficult task. Complicating the rescue is the size and weight of the victim compared with that of the fire fighter.

> **Key Points**
>
> Ladders used for ventilation at windows should be positioned alongside the window to the windward side with the tip about even with the top of the window. In this way, the fire fighter is positioned away from the front of the window when it is broken.

Ventilation

Ground ladders can be placed to provide venting crews with access to roofs or windows. Ladders used for ventilation at windows should be positioned alongside the window to the windward side with the tip about even with the top of the window. In this way, the fire fighter is positioned away from the front of the window when it is broken. As noted, if a ground ladder is raised to a flat roof, the tip should extend four to five rungs above the roof line for ease in climbing on and off the roof. It also allows fire fighters to locate the ladder when it is time to vacate the roof.

A roof ladder is used to gain access to a sloped or pitched roof. A roof ladder can be placed on the roof using either one or two fire fighters. The object is to slide the roof ladder over the peak of the roof and secure the ladder with hooks attached at the tip. This allows the fire fighter's weight to be distributed evenly over the roof area and gives a secure platform to work from.

Advancing Hose Lines

A number of situations may require the use of ground ladders to advance hose lines into a building. For example, occupants might be leaving by stairways at the same time engine companies are advancing attack lines or positioning hose lines to protect escaping victims. If the stairways are not well located, engine crews may use ground ladders to get into advantageous

> **Key Points**
>
> A roof ladder can be placed on the roof using either one or two fire fighters. The object is to slide the roof ladder over the peak of the roof and secure the ladder with hooks attached at the tip. This allows the fire fighter's weight to be distributed evenly over the roof area and gives a secure platform to work from.

> **Key Points**
>
> A number of situations require the use of ground ladders to advance hose lines into a building.

positions to attack the fire, as shown in **Figure 9-6**. Sometimes there are no stairways at all at the rear or sides of a building. Also, stairways, although free of people, might be damaged by fire or explosion.

When interior stairways and corridors are available, they will normally be used to advance initial attack lines to ensure control of these passageways. If additional hose lines are advanced through the same stairways, they could become overcrowded and thus unsafe for fire fighters. Ground ladders provide an obvious means of relieving such conditions.

Placing Ladders

Ground ladders should be placed where they will be of use in the overall fireground operation such as advancing initial attack lines, backup lines, or hose lines that might be used to cover interior and exterior exposures. The initial sizeup by ladder company officers will indicate where engine companies may need to enter the building. Ladders should be raised at these positions on arrival; that is, ladder crews should raise the ground ladders at strategic points as part of a coordinated fire attack.

Ground ladders should not be raised simply as an exercise, as this could delay other important ladder company operations. It is wasteful to raise a ground ladder where it obviously will not be needed, but it is just as wasteful not to raise a ladder where it might be needed. Occasionally, a properly positioned ground ladder will not be needed by engine crews, possibly because the fire was quickly controlled from other positions, but this should not deter ladder companies from placing ground ladders in similar positions at future fires where the ladders could be crucial.

After ladders have been raised for rescue, forcible entry, ventilation, and hose advancement, a detached building can be laddered. An attached building may be laddered only at the front

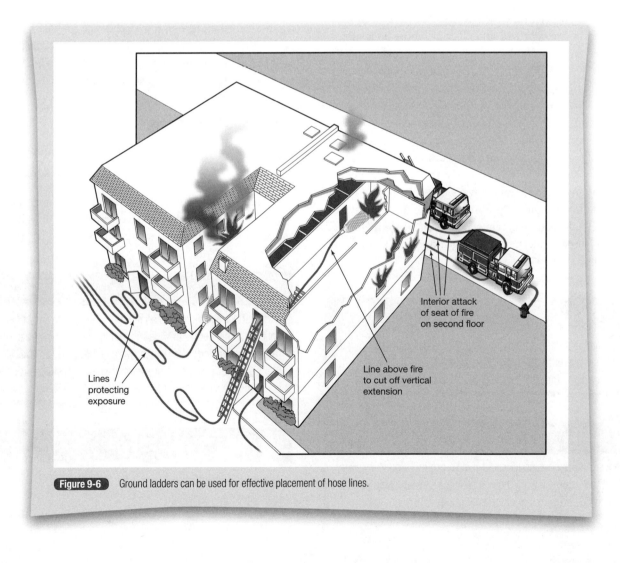

Interior attack
of seat of fire
on second floor

Line above fire
to cut off vertical
extension

Lines
protecting
exposure

Figure 9-6   Ground ladders can be used for effective placement of hose lines.

Key Points

If at all possible, ladder crews should avoid placing ladders in front of building entrances, where fire fighters entering the building or occupants leaving it could inadvertently move it out of a safe position or knock it down.

and rear. If at all possible, ladder crews should avoid placing ladders in front of building entrances, where fire fighters entering the building or occupants leaving it could inadvertently move it out of a safe position or knock it down. If hand lines become tangled in the butt, attempts to free them will delay fire attack and could require the ladder to be repositioned.

Climbing Ladders with a Hose Line

Carrying a hose line up a ladder is best accomplished when the line is not charged. Normally, in climbing ground ladders, one fire fighter should be on each section of the ladder to prevent overloading. If the ladder was raised at a less-than-acceptable angle, this distance should be increased or the ladder repositioned for the proper climbing angle. The lead fire fighter drapes the hose line over the shoulder and advances the line up the ladder to the first fly section. A second fire fighter drapes a large loop of hose on his or her shoulder and starts to climb the ladder. If a third fire fighter is needed, the process is repeated as the hose line is advanced up each section of ladder. As the fire fighters climb, they should allow loops of hose to hang off the side of the ladder. This will distribute the total weight of the hose properly along the ladder. A fire fighter should be located on the ground to assist in the movement of hose up the ladder. A fire fighter should also heel or secure the butt of the ladder to prevent movement. The hose line should be charged when it has been advanced into the building and located in the area of use.

Positioning Firefighting Streams

Ground ladders can be used to position and operate hose streams being directed into a building. There are two methods of placing ladders for this purpose. In the first, the topmost three or four rungs are allowed to extend over the sill and into the window. The hose line is then carried up and tied to the ladder so that the stream will be directed through the window opening. The nozzle is opened and the stream directed onto the fire, as shown in **Figure 9-7**.

Figure 9-7 The ground ladder is placed so that the top three rungs extend into the window. The hose line is carried up and tied to the ladder so that the stream will be directed through the window onto the fire.

In the second method, the ladder is raised over the window with the tip placed against the wall above it. Sometimes the tip is placed on the sill or into the next higher window. Again, a hose line is carried up, tied to the ladder, and directed onto the fire.

In either case, the ladder should be tied in to the building to prevent movement. If the ladder is in a window, a tie-in with a rope hose tool or dog chain will suffice. If the ladder is against a wall, a pike pole can be placed across the inside of a lower window, tied to a rung, and pulled tight against the inside wall. The ladder should be braced by fire fighters at the butt by

Key Points

Normally, in climbing ground ladders, one fire fighter should be on each section of the ladder to prevent overloading.

Key Points

Ground ladders can be used to position and operate hose streams being directed into a building. There are two methods of placing ladders for this purpose.

either heeling the ladder or using another method of securing the butt.

The hose line should be tied to the ladder with a rope hose tool or strap in a manner that will allow the rope to absorb some of the reaction of the hose line when the nozzle is opened. It should, therefore, not be tied rigidly to ladder rungs; rather, it should be suspended between rungs by the rope or strap. One way to do this is to hold the nozzle between two rungs, pass the rope behind the rung just above the nozzle, wrap it around the rung to take up the slack, and then hook it to the second rung up. With the hose suspended from the rope, it can move back when the nozzle is opened, but it will not exert a strong pull on the ladder. The ladder will remain in place.

After the stream has knocked down the fire in an area, it can be shut down, untied, advanced, recharged, and used for interior fire attack; however, fire fighters must use care after the stream is shut down. Any steam or smoke in the area should be allowed to clear; then the condition of the interior, especially the floor, should be checked. Before they enter the building, fire fighters should make sure that the fire is extinguished and has not flared up again after the hose line was shut down.

If it is safe to enter the building through the window, the hose line should be untied from the ladder and placed over the sill. The nozzle operator should enter the window only when a second fire fighter has climbed to a position immediately outside the window to assist in advancing the hose line and back up the nozzle operator if fire erupts in the area.

Fire fighters working from a ladder should secure themselves to that ladder by using a leg lock or approved safety harness or belt. Hoisting a ladder by rope to where it will be used may be much quicker than trying to maneuver a fairly long ladder up several flights of stairs.

> ## Key Points
>
> If it is safe to enter the building through the window, the hose line should be untied from the ladder and placed over the sill. The nozzle operator should enter the window only when a second fire fighter has climbed to a position immediately outside the window to assist in advancing the hose line and back up the nozzle operator if fire erupts in the area.

> ## Key Points
>
> Hoisting a ladder by rope to where it will be used may be much quicker than trying to maneuver a fairly long ladder up several flights of stairs.

> ## Key Points
>
> When a ladder is being hoisted with a rope, the rope should be tied several rungs below the top. When the rope is pulled up, these rungs will extend above the windowsill or rooftop, allowing ladder crews to get a good hold on the ladder and use some leverage to get it in to the building or on to a roof.

When a ladder is being hoisted with a rope, the rope should be tied several rungs below the top. When the rope is pulled up, these rungs will extend above the windowsill or rooftop, allowing ladder crews to get a good hold on the ladder and use some leverage to get it in to the building or on to a roof. This is a formidable task, and those participating in this task must coordinate their efforts to accomplish the operation safely.

Bridging

Under no circumstance should a ground ladder be used horizontally to bridge two buildings from a window, roof, balcony, or other platform over an alley, walkway, narrow court, or in an air shaft or elevator shaft.

Raising Ladders from Roofs

Ground ladders can be raised to the sides of a tall building from the roof of an attached or nearby lower building. Ladder crews should first raise a ladder to the roof of the lower building, climb up to the roof, and then get additional ladders up to that roof. If the lower building is not more than three stories high, additional ladders can be lifted up by crews on the ground to ladder company members on the roof, who then should be able to get hold of the ladders and pull them up. If the lower building

> ## Key Points
>
> Under no circumstance should a ground ladder be used horizontally to bridge two buildings from a window, roof, balcony, or other platform over an alley, walkway, narrow court, or in an air shaft or elevator shaft.

> ## Key Points
>
> Ground ladders can be raised to the sides of a tall building from the roof of an attached or nearby lower building.

is more than three stories high, the ladders should be hoisted up by rope. After additional ladders are on the roof of the lower building, they are raised to the proper positions on the side of the taller building.

Covering Weakened Areas

Ground ladders sometimes are placed over building features weakened by fire or by firefighting operations to allow safe passage. For example, a stairway damaged by fire might still be used if a ground ladder is laid over the stairs, with the butt and tip in contact with firm flooring or undamaged stairway supports.

A ground ladder also can be placed over a weakened or suspect area of a roof or floor to allow crews to work safely in that area. Each end of the ladder must be supported by a solid part of the roof or floor.

Wrap-Up

Chief Concepts

- Ground ladders serve many purposes in firefighting operations. Regardless of how the ladder is used, it must be handled safely in all facets of its operation. A ladder must be carried from the apparatus, spotted, raised, pivoted if necessary, and extended. The ladder must then be climbed to perform the firefighting tasks of rescue, ventilation, forcible entry, and fire attack. The proper ladder must be selected for the job it has been designated to do. The weight and length of each ladder will dictate how many fire fighters it will require to place it in service on the fire ground.

- Training with ground ladders is essential if ladder company personnel are to perform efficiently on the fire ground. In training sessions, ladders should be handled by the same number of fire fighters as will be available on the fire ground and not by larger than normal crews. When help will be required to raise the larger extension ladders, for example, 40- to 50-foot ladders with staypoles, engine personnel should be trained to assist in this operation.

Key Terms

Combination ladders: Can be used as a single ladder, an extension ladder, and an A-frame.

Extension ladders: Adjustable in height and has a base ladder and one or more fly sections that travel in guides allowing for extension of the fly section.

Folding ladders: Single ladder that has hinged rungs; this allows the ladder to be folded so that the beams rest against each other.

Pompier ladders: Consist of a single-beam ladder with rungs protruding from either side. A large hook at the top is placed on the sill in an open window or other opening. Used to climb from floor to floor by way of the window openings.

Roof ladders: Designed in one section only, equipped with folding hooks at the tip to provide an anchoring mechanism when the ladder is placed over the peak or ridgepole of a building. Designed to lie flat on a roof as a work platform.

1. The main advantage of a ground ladder over an aerial ladder or elevating platform is
   a. Its portability
   b. Its reach
   c. It takes fewer fire fighters to raise
   d. It is more secure and stable

2. Which of the following is the most popular type of ground ladder in the fire service?
   a. Aluminum alloy
   b. Steel alloy
   c. Fiberglass
   d. Wood

3. One or more fly sections would be found on a(n)
   a. Single ladder
   b. Wall ladder
   c. Roof ladder
   d. Extension ladder

4. Which type of ladder carry is used when encountering a narrow opening?
   a. Flat shoulder
   b. Flat arms length
   c. Overhead
   d. Underarm

5. If an extension ladder needs to be pivoted, the pivot should be done
   a. After the ladder is extended to reach the target
   b. Before the ladder is extended to reach the target
   c. After the ladder is raised when performing a flat raise but after the ladder is raised for a beam raise
   d. After the ladder is raised when performing a beam raise but after the ladder is raised for a flat raise

6. A minimum of three fire fighters is usually required when raising an extension ladder that is _____ feet or longer.
   a. 24
   b. 30
   c. 35
   d. 40

7. The NFPA *Standard on Use, Maintenance, and Service Testing of In-Service Fire Department Ground Ladders* is
   a. NFPA 1500
   b. NFPA 1710
   c. NFPA 1901
   d. NFPA 1932

8. A simple formula used to determine the correct placement of the butt from the base of the building is
   a. Multiply the length of the ladder by five then add two
   b. Multiply the working length of the ladder by five then divide by two
   c. Position the bottom of the ladder at a distance from the vertical plane equal to one fourth the total working length of the ladder
   c. Divide the working length of the ladder by four

9. According to NFPA standards, the maximum load (persons, equipment, and any other weight) shall not exceed _____ pounds on a combination ladder.
   a. 300
   b. 500
   c. 750
   d. 1,000

10. When rescuing occupants from a window using a ground ladder, the tip of the ladder should _____
    a. Be placed to the side of the window and even with the windowsill
    b. Be placed to the side of the window and at least three rungs above the windowsill
    c. Be placed inside the window so at least three rungs are above the windowsill if the window is large enough to accommodate the ladder and people being rescued at the same time
    d. Usually be at or just over or above the level of the windowsill

11. Ground ladders raised to provide alternative egress are best placed at
    a. Fire escape balconies
    b. Porches or apartment balconies
    c. Windows
    d. Windows leading to corridors

12. Of the methods listed below, the *least desirable* way to remove an unconscious victim is
    a. The interior stairs
    b. The fire escape
    c. Via ground ladder
    d. Via an aerial ladder

13. Ground ladders can be used to horizontally bridge
    a. From the window of one building to another
    b. From the balcony of one building to another
    c. From the roof of one building to another
    d. Under no circumstances should a ground ladder be used to horizontally bridge between two buildings

Property Conservation

Chapter 10

Learning Objectives

- Define property conservation.

- Consider the time frame in which property conservation should begin.

- Identify the equipment and procedures used to protect the building contents.

- Identify two ways of performing property conservation by removing water from a building.

- Identify the methods of removing smoke from within a building.

Property-conservation and overhaul operations help reduce property damage to both the building and its contents. Property conservation should begin during firefighting operations. Fire fighters who are engaged in this operation should be aware of fire growth or structural damage to the building. If conditions deteriorate or if a defensive operation is to be conducted, fire fighters should not be placed in jeopardy attempting to save property.

Property-conservation operations may begin with a limited number of personnel because of search-and-rescue and fire-suppression operations being conducted at the same time. The incident commander must realize that adequate staffing levels will be needed to conduct property-conservation operations safely and efficiently. Fire fighters must be in full personal protective clothing, including SCBA, to perform the job safely.

In addition to covering or removing furnishings and equipment, reducing water damage and the accumulation of water from the building will alleviate added weight to the floors of a structure already weakened by fire.

Property conservation is the third tactical priority behind life safety and extinguishment. Fire fighters perform property conservation to minimize the damage that water and smoke cause during a fire.

In recent years, property conservation has been referred to as the "lost art" of the fire service. In many cases, property conservation is neglected because of a supposed lack of personnel. Command must make tactical decisions that will ensure that all firefighting priorities are accomplished on the fire ground. If property-conservation efforts must be addressed and the number of personnel on scene is inadequate, command should request additional resources. If additional resources are not available, property conservation may have to be delayed until life safety and extinguishment priorities have been completed or are under control. Crews may have to begin property-conservation efforts simultaneously when staffing levels are inadequate.

Unfortunately, after a fire department or an incident commander believes that adequate resources are not available for effective property-conservation efforts, that work will not be done, regardless of the actual personnel situation. There are fire departments that make little or no provisions for property-conservation efforts, possibly believing that there is little time to devote to this tactical priority during firefighting.

It makes little sense for fire fighters to put all their effort into controlling a fire that may do a few hundred dollars' worth of damage while allowing water to ruin thousands of dollars' worth of office equipment, domestic furnishings, or other valuable property and irreplaceable contents—yet that is often the case.

Another detriment to property-conservation operations is the misconception that it is related to overhaul. This idea is perpetuated by such outdated phrases as salvage and overhaul, whose very nomenclature indicates a connection between the two operations. Actually, they are no more related than any other two ladder company operations on the fire ground. They have different objectives, are initiated at different times, and require different procedures.

The main objective of property conservation is to protect the building and its contents from unnecessary water and smoke damage; the main objective of overhaul is to make sure that the fire is completely out. Although these two procedures may take place at the same time, they are different duties and require different skills.

Property conservation should be started as soon as possible after fire attack begins; overhaul operations begin when the fire is under control. Property conservation is performed with equipment designed for the task, including **salvage covers, conduits,** and chutes to channel water, submersible and portable pumps, water vacuums, venturi siphons, drain screens, squeegees, mops, brooms, PPV fans and blowers, and associated equipment; overhaul requires truck tools and hose lines.

The command staff must also understand that if fire conditions are deteriorating or if a defensive mode of operations is being conducted, fire fighters should not be placed in jeopardy attempting to save property. Fire fighters should never be placed in grave danger in an attempt to save property.

This chapter covers property-conservation activities while overhaul operations are discussed in Chapter 12. It is important that the command staff as well as ladder crews be thoroughly

Key Points

The main objective of property conservation is to protect the building and its contents from unnecessary water and smoke damage; the main objective of overhaul is to make sure that the fire is completely out.

Key Points

If property-conservation efforts must be addressed and the number of personnel on scene is inadequate, command should request additional resources.

Key Points

Fire fighters should never be placed in grave danger in an attempt to save property.

trained in both operations so that they understand the differences and do not neglect either operation on the fire ground.

Water and Property Conservation

Water is the prime extinguishing agent that fire companies use, and it will probably remain so for some time to come. Fire departments are organized and trained to apply water onto and into structures. Hose streams are needed to protect victims until they can be rescued, to protect exposures, and to contain and extinguish fires; however, the same water used to limit the damage done by fire can itself damage a building structurally and ruin its contents. Property-conservation efforts can reduce water damage in almost any structure or occupancy.

There are two ways of performing property-conservation operations when water is involved. One method protects the contents of a building, and the other protects the building itself from structural damage due to the weight of the water. The first method, which is shown in **Figure 10-1a**, mainly requires the proper placement of salvage covers to protect the contents and capture and/or displace the water. The second approach, which is shown in **Figure 10-1b**, requires that the water be removed from the building before the weight of the water weakens the structural stability of the structure. Both methods are equally important in reducing property losses.

If the building is equipped with a standpipe system, the control valve should be located and the system shut down after the fire has been contained and is under control. A fire fighter should be stationed at the control valve in case the system needs to be turned back on. Individual sprinkler heads can be shut off by the use of sprinkler stops.

Property-conservation efforts should include capturing water run off before it has a chance to migrate to another location in the building. This can be accomplished by using salvage covers

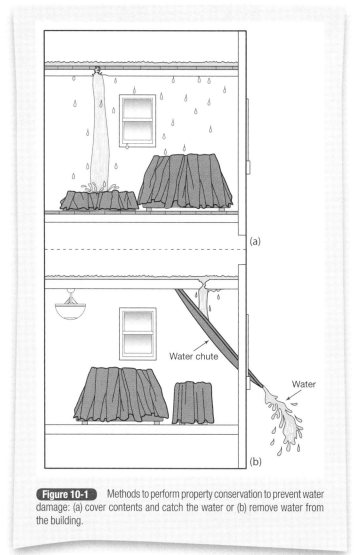

Figure 10-1   Methods to perform property conservation to prevent water damage: (a) cover contents and catch the water or (b) remove water from the building.

as **catchalls.** Catchalls can be rigged to capture and hold either small or large amounts of water in a particular area.

Channel the water out of the building by using chutes or other improvised devices. The water can be moved out through stairways, windows, or drains in the floors. Squeegees, brooms, and mops can be used to direct the water along the passageway and out of the building.

Salvage covers must cover valuable property early in the incident. There is no sense in providing property-conservation efforts after the property has been destroyed by water. As noted

Key Points

The same water used to limit the damage done by fire can itself damage a building structurally and ruin its contents. Property conservation can reduce water damage in almost any structure or occupancy.

Key Points

There are two ways of performing property-conservation operations when water is involved. One method protects the contents of a building, and the other protects the building itself from structural damage because of the weight of the water.

Key Points

Property-conservation efforts should include capturing water run off before it has a chance to migrate to another location in the building. This can be accomplished by using salvage covers as catchalls.

Key Points

Salvage covers must cover valuable property early in the incident.

earlier, command must make a concerted effort to get property-conservation efforts underway. As soon as possible, valuable property should be grouped together and covered with salvage covers.

If there is an adequate number of personnel on scene, property could be moved from an area where water is creating a problem to another area inside the building. On occasion, the property may need to be removed from the building, as shown in **Figure 10-2**. This is a labor-intensive endeavor and may take some time to complete depending on the scope of the operation. After the property has been removed to the outside, it must be protected from the elements, theft, and vandalism.

The type of property-conservation operations required in a particular building will depend to some extent on its construction. For example, the floors in a fire-resistant building are made of concrete, which usually will hold the water directed onto it. Property conservation in such a building primarily requires containing the water, directing it along preset paths, and removing it through floor drains, windows, stairways, or chutes; however, some water might seep down to lower floors through cracks in the cement floors, where pipes or other

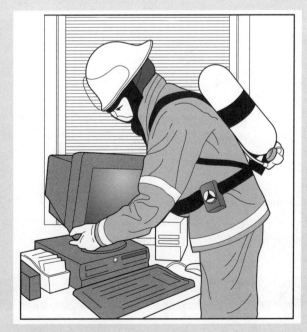

Figure 10-2   Valuable property is sometimes removed from the building by a fire fighter to save it from water damage.

Key Points

The type of property-conservation operations required in a particular building will depend to some extent on its construction.

utilities pass through the floor, and through stairways (see **Figure 10-3**). Therefore, the contents of lower floors also must be protected.

Water will seep quickly through wooden floors, as shown in **Figure 10-4**. Because of this, the first property-conservation measure must be directed toward protecting the contents of the floor below the fire floor. This should be accomplished by using salvage covers; however, it is also necessary to remove the water from such buildings to keep it from seeping from floor to floor.

Preplan and sizeup of the fire situation should indicate which property-conservation measures should be initiated first and where. To be effective, property conservation should begin along with the fire attack or as soon as possible thereafter; that is, the building and its contents should be protected from water damage when water is first directed into the building. There should be no hesitation in calling for extra companies to perform property conservation when it is obvious that uncontrolled water damage could be extensive.

Protecting the Building Contents

Building contents are primarily protected by covering them to keep water and debris from damaging them. The flow of excess water should be directed away from stock, furnishings, and equipment. Valuable contents that can be lifted off the floor may be placed on less-valuable contents or those that cannot be moved. A salvage cover should then cover them. Also, catchalls can be fashioned to catch water dripping

Key Points

To be effective, property conservation should begin along with the fire attack or as soon as possible thereafter.

Key Points

Valuable contents that can be lifted off the floor may be placed on less-valuable contents or those that cannot be moved. A salvage cover should then cover them.

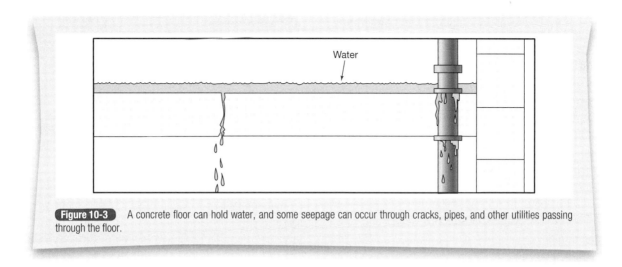

Figure 10-3 A concrete floor can hold water, and some seepage can occur through cracks, pipes, and other utilities passing through the floor.

Figure 10-4 Wooden floors are generally porous, which allows water to seep through them quickly.

from above. Salvage covers are an important item in all three operations.

Salvage Covers

Salvage covers are essentially large sheets of waterproof material. They are available in several materials, dimensions, and shapes. Some are fire resistant, and some are not. Some salvage covers must be placed with a particular side open to the water. Fire fighters should be familiar with the type of salvage covers carried by their company, as the type often determines how the covers can be used. It is equally important that salvage covers be purchased with regard for the average number of personnel responding with the company. The size and weight of a cover can affect the number of people required to place it. Most fire departments now carry plastic sheets that are used as salvage covers. In the past, many departments used heavy canvas tarps as salvage covers. Because of the cost, maintenance, and difficulty of use, most departments now have shifted to the low cost and disposable lightweight plastic sheets. They are made in a variety of colors and weight. Fire departments should select salvage covers that are able to be reused and stand up to the strains and use during fireground operations.

Various ways of folding salvage covers for transporting them on the apparatus are given in various manipulative-skill manuals. The appearance of the folded cover should indicate whether one fire fighter can spread it or whether two or more fire fighters are required. If it does not, then the cover should be tagged to indicate its size and the number of personnel required for its use and any special use for which it has been designed or folded.

The types of salvage covers used, the folding methods, and the carrying and spreading techniques should be standardized within a department and, if possible, with neighboring fire departments. This will allow covers to be used interchangeably and spread quickly so that crews providing property conservation can be released for other duties as required by the fire situation.

Key Points

Fire fighters should be familiar with the type of salvage covers carried by their company, as the type often determines how the covers can be used.

Key Points

The types of salvage covers used, the folding methods, and the carrying and spreading techniques should be standardized within a department and, if possible, with neighboring fire departments.

Key Points

Salvage covers should first be spread over the building contents that are in the most danger of being damaged by water. In most cases, these are on the floor below the fire floor, directly under the fire; that is, under the area to which the streams will be directed.

Covering the Building Contents

Salvage covers should first be spread over the building contents that are in the most danger of being damaged by water. In most cases, these are on the floor below the fire floor, directly under the fire; that is, under the area to which the streams will be directed. These contents will be subjected to water seeping from above. After the contents of this area are well protected, covers should be placed on the contents of surrounding areas.

In some cases, the area under the fire may contain few items that would be damaged by water. In this instance, either the contents of surrounding areas should be covered first, or if the fire floor contains equipment or furnishings that are sensitive to water damage, its contents should be covered first. The important point is to first cover the items that could suffer the most water damage because of either their position or their value.

Besides being spread to protect items resting on the floor, salvage covers can be rigged over shelves mounted on walls, as shown in **Figure 10-5**. They can be nailed to the wall above the shelves through grommets provided in the covers and draped to cover the shelves. For this use, hammer-and-nail kits should be part of the equipment that ladder companies carry.

Salvage covers should be spread over the contents of the fire floor and floors above the fire, when necessitated by firefighting operations in those areas. For instance, before a ceiling is pulled down to search for the lateral spread of fire, covers should be placed over nearby items to keep debris and water, if used, from damaging them. The fire situation and the available personnel will determine whether such property-conservation measures are feasible. Crews entering an area with pike poles to pull ceilings can easily carry and spread covers if the fire situation is not too severe. When covers are not available, unnecessary damage can be avoided by moving the room's contents away from the work area.

When the number of salvage covers is limited, the available covers should be used to protect the most valuable contents. Every salvage cover should be used as effectively as possible.

Key Points

When covers are not available, unnecessary damage can be avoided by moving the room's contents away from the work area.

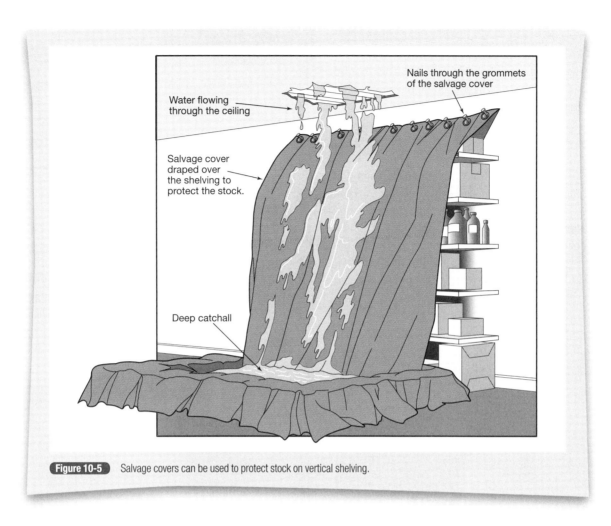

Water flowing
through the ceiling

Nails through the grommets
of the salvage cover

Salvage cover
draped over
the shelving to
protect the stock.

Deep catchall

Figure 10-5 Salvage covers can be used to protect stock on vertical shelving.

One way to do this is to move the contents into as small a space as possible before covering them. If the contents of an average room of a dwelling are piled together, they can usually be fully protected with two or three salvage covers.

Controlling the Water Flow

Salvage covers can be used to control the flow of water along a floor. Placed in doorways or stuffed under doors, they will block the movement of water out of the fire area or the area below the fire. In the latter case, the contents should previously have been protected with salvage covers; however, water that is blocked and left to accumulate will soon begin to leak down through the building. It must therefore be removed down stairways or other openings. These operations are discussed in greater detail in the next section.

Salvage covers can be folded for use as conduits to direct accumulated water to stairways and then down the stairways and out through exterior doorways. The water could also be directed to a balcony or porch and out over the side of the building. Salvage covers to be used as conduits should be prefolded at the station, rolled up, tied, and carried on the apparatus ready for use. This saves time that would otherwise be needed to rig them on the fire ground.

To form a conduit, two salvage covers are laid out flat, end to end, with some overlap. The sides of both covers are then rolled

Key Points

When the number of salvage covers is limited, the available covers should be used to protect the most valuable contents.

Key Points

Salvage covers can be used to control the flow of water along a floor. Placed in doorways or stuffed under doors, they will block the movement of water out of the fire area or the area below the fire.

Key Points

Salvage covers can be folded for use as conduits to direct accumulated water to stairways and then down the stairways and out through exterior doorways.

together toward the center to form one narrow length. This is then coiled up tightly from one end and tied securely.

To use the conduit, fire fighters carry it to the top of the stairs, unfasten it, unroll it down the stairs, and spread the sides apart. The rolls along the edges keep the water within the conduit. Water can be guided to the conduit with squeegees and brooms or through a second conduit, if required, placed between the accumulated water and the stairway.

Conduits are also used in two ways to guide water along floors. On watertight floors, they can be unrolled to form a guide or wall to keep the water confined as it is moved toward stairs or a drainage system, as shown in **Figure 10-6**. A pair of conduits used in this manner can be very effective. On porous floors or those that are not watertight, the conduit is unrolled and spread flat, except for a roll on each side to hold water. The water is moved over the conduit in the desired direction.

Care must be taken when conduits are made up and placed to make sure that the interlock works effectively. The cover forming the upper part of the interlock must be placed toward

Key Points

Care must be taken when conduits are made up and placed to make sure that the interlock works effectively.

the flow of water ("upstream") so that the water will pass over it and down onto the next cover.

Catchalls

Salvage covers can be rigged as basins, generally referred to as catchalls, to catch and hold water dripping from overhead, as from a ceiling. The cover can be rolled in quickly from all edges to form a flat, shallow catchall, as shown in **Figure 10-7**, or draped over four ladders, inflated fire hose or other suitable material to form a deep catchall, as shown in **Figure 10-8**. Because they are small, catchalls cannot be used as the sole water damage control device when extensive firefighting operations are in progress or

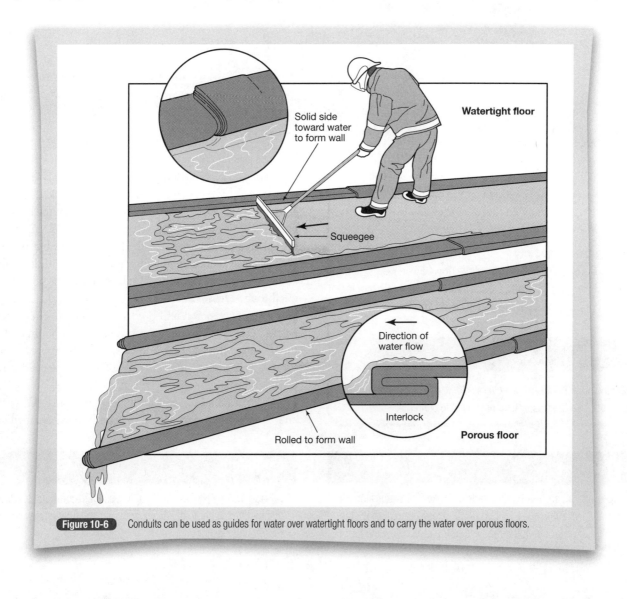

Figure 10-6   Conduits can be used as guides for water over watertight floors and to carry the water over porous floors.

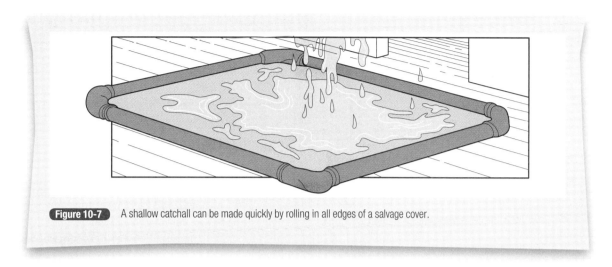

Figure 10-7 A shallow catchall can be made quickly by rolling in all edges of a salvage cover.

Key Points

Salvage covers can be rigged as basins, generally referred to as catchalls, to catch and hold water dripping from overhead, as from a ceiling.

large amounts of water, as may be encountered with a broken pipe, are involved. They are most effective at comparatively small fires or water incidents.

Catchalls are, however, effective in keeping moderate amounts of water off building contents and in preventing water from moving across the floor or seeping down to lower floors.

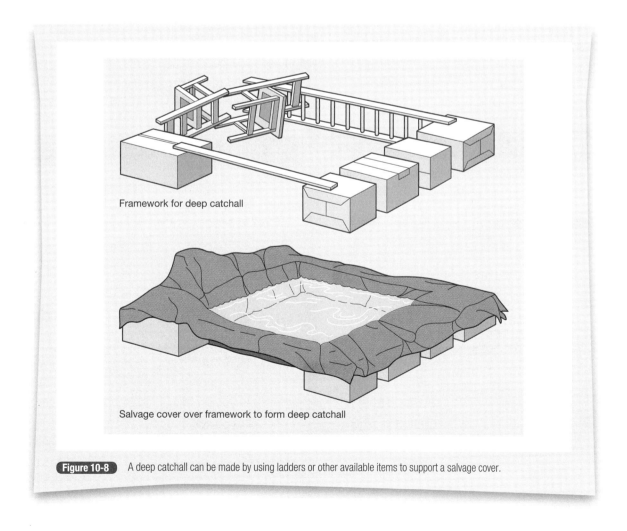

Framework for deep catchall

Salvage cover over framework to form deep catchall

Figure 10-8 A deep catchall can be made by using ladders or other available items to support a salvage cover.

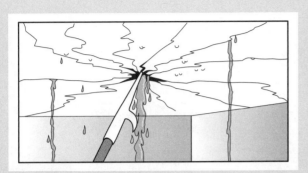

Figure 10-9 Converging cracks in the ceiling indicate an accumulation of water. To relieve the water, a small hole should be poked gently where the cracks converge.

Often fire fighters use pike poles to punch holes in the ceiling where the water is dripping, as shown in **Figure 10-9** . This helps drain water and prevent the collapse of a big piece of the ceiling, which would splash water over a large area. After the catchall is filled during these operations, it must be dumped carefully or drained using a pump or water vacuum so that the water does not spill over onto the floor and pass through to lower stories.

If the full catchall is too heavy to move, portable pumps can be used to remove the water. Care should be taken to make sure the connections on the hose lines carrying water out of the building are tight and do not leak. This is yet another way of providing property conservation.

Key Points

Catchalls are effective in keeping moderate amounts of water off building contents and in preventing water from moving across the floor or seeping down to lower floors.

Key Points

If the full catchall is too heavy to move, portable pumps can be used to remove the water.

Removing Water from Buildings

A large volume of water, possibly weighing several tons, can collect within a structure while an extensive fire is being fought **Table 10-1** . Although salvage covers might be placed below and around the fire, collected water will run down through the structure. This damages uncovered contents of lower floors becomes deep enough in spots to damage items resting on the floor and, most important, will add weight to the floors of a structure already weakened by fire. Ladder company personnel must remove water before it can do further damage or cause the building to collapse.

The sooner the water is removed, the less damage it will do to floors and carpets and the less chance there is of leakage to lower floors. Often, there will be little or no permanent damage to flooring materials or to the ceilings below these floors if the water is removed quickly and the floors allowed to dry.

One particularly effective way to remove water from smaller buildings is to quickly push it out with squeegees or brooms. Where the water is to be pushed through a hallway, salvage covers should be stuffed under doors and placed along wide openings to rooms to keep the water from entering dry areas, as shown in

Key Points

Ladder company personnel must remove water before it can do further damage or cause the building to collapse.

Table 10-1   Weight of Water from Master Streams

Number of Master Streams at 1,000 gpm (3,785 L/min) each	Weight Added Each Minute at 8.33 lb/gal (1 kg/L)	Total Weight at 30 Minutes lb (kg)	Total Weight at 60 Minutes lb (kg)
1	8,330 lb	249,900 lb	499,800 lb
	(3,785 kg)	(113,550 kg)	(227, 100 kg)
3	24,990 lb	749,700 lb	1,499,400 lb
	(11,355 kg)	(340, 650 kg)	(681,300 kg)
10	83,300 lb	2,499,000 lb	4,998,000 lb
	(37,850 kg)	(1,135,500 kg)	(2,271,000 kg)

Source: *Structural Fire Fighting*, Table 11.1, p. 326, NFPA: Quincy, MA, 2000.

Figure 10-10 Salvage covers help block water from flowing into dry areas.

Key Points

More permanent chutes can be made from salvage cover material. These chutes can be used to direct water from one floor out through the windows or other openings of the floor below.

used to direct water from one floor out through the windows or other openings of the floor below.

Each chute is made from a strip of salvage cover material about 10 to 12 feet long and a pair of wooden poles or aluminum pipes of the same length. The long edges of the strip are rolled and fastened around the poles. The chute is rolled pole to pole into a compact unit for transportation. For use, it is unrolled, rigged below a hole in the floor, and placed so that it angles down to a window on the next lower story. The chute allows the water to escape directly from the floor above with minimum movement through the building.

The chute can be rigged below the hole on a short straight ladder or an A-frame type ladder. The ladder should be tall enough to hold the upper end of the chute close enough to the hole to keep the water from splashing down. The chute poles can be tied to the ladder with short pieces of rope provided on the chute. Salvage covers should be spread over items close to the chute to prevent damage if there is a spill.

The lower end of the chute should extend far enough out the window so no water will fall back into the building. Windows below should be closed or covered if broken. Personnel must be warned to stay clear of the area.

Another effective chute can be fabricated from a funnel-type device and a length of fire hose. One chute of this type is illustrated in **Figure 10-11**. To use this chute, ladder crews cut a hole in the floor (the water is blocked with a salvage cover while the hole is being cut). A bar is laid across the hole to support the chute, and the chute is positioned. The hose is run out a window

Figure 10-10. When this operation is started early enough, there should be little or no leakage to lower floors. The effectiveness of the operation also depends on the flooring material and the quality of its construction.

A number of other ways to remove accumulated water from a building are discussed in the remainder of this chapter. Most are more complex than the squeegee operation, but they may be necessitated by the construction of the building or by the amount and location of water to be removed. Usually, a combination of several methods will be required to "dry out" a building.

Chutes

The use of rolled salvage cover conduits and makeshift devices has already been discussed. In addition, more permanent chutes can be made from salvage cover material. These chutes can be

Key Points

Salvage covers should be spread over items close to the chute to prevent damage if there is a spill.

Key Points

One particularly effective way to remove water from smaller buildings is to quickly push it out with squeegees or brooms. Where the water is to be pushed through a hallway, stuff salvage covers under doors and place them along wide openings to rooms to keep the water from entering dry areas.

Key Points

The lower end of the chute should extend far enough out the window so no water will fall back into the building. Windows below should be closed or covered if broken. Personnel should be warned to stay clear of the area.

Figure 10-11   Chutes can be used to remove water from a building quickly and efficiently.

or into the drainage system. The salvage cover is then arranged to form a U-shaped dike to direct the water toward and into the hole. The water can be pushed to the hole with squeegees. A drain screen can be constructed as part of the top of the chute or placed in the chute when it is positioned in the hole. Although funnels are seldom used in this manner, they do work well and can be set up in a relatively short period of time if they are part of a ladder company's equipment.

Drains

Floor and wall drains built into the building can be used if they are located fairly close to the accumulations of water. Water should not be moved long distances across floors if this can be avoided. In any case, large quantities of water should not be moved to small drains because, if the capacity of a drain is exceeded, another accumulation of water will be formed. Also, the drain must be kept free of debris so that water does not back up in the area. In most situations, built-in drains must be used in combination with other water removal methods. Locations of adequate drains should be determined during preincident planning surveys.

Floor drains might be located throughout a building or only in certain areas. They usually have some sort of a screen-type cover. Some drain directly into sewers and others into dry wells. They are ideal outlets for water because they are designed for that purpose. Little effort is involved in their use, except to direct water to them and keep the covers free of debris.

In some buildings, wall drains (scuppers) have been placed at the base of the walls even with the floor. The interior opening is usually covered with a heavy wire mesh; the exterior opening usually has a hinged cover. Scuppers are also ideal for removing water and are used in the same way as floor drains.

Toilets

When a toilet is unbolted from the floor and lifted out of place, a waste pipe opening, connected to the sewer system, is exposed. In most cases, the opening is 4 inches or more in diameter and can be used to remove much water quickly from nearby areas if the drain is kept from clogging. The flooring in bathrooms is usually the most water resistant in a building, and thus, this operation can be very effective. It should be repeated as necessary, from story to story, when water has accumulated throughout a building.

Key Points

An effective chute can be fabricated from a funnel-type device and a length of fire hose.

Key Points

In most situations, built-in drains must be used in combination with other water removal methods. Locations of adequate drains should be determined during preincident planning surveys.

The waste pipe may not be sealed to the floor around it so that water can drop down to lower stories through the space between the pipe and the floor. To prevent this, the space should be closed up as much as possible. Anything that can be stuffed around the pipe will hinder the flow of water into the opening. A salvage cover can be placed on the far side of the pipe to keep water from flowing past it, and part of the cover could be stuffed in around the pipe. A towel or cloth may also be used. Ceilings below the toilet should be checked for leakage.

To prevent the waste pipe from becoming clogged with debris, a drain screen should be placed over the opening. An effective screen can be made up of 1-inch mesh on a 10- to 12-inch steel ring or square frame.

Fire departments should be able to purchase or fabricate items such as mesh screens, funnels or funnel-type devices, framework to form deep catchalls, or other ways of collecting and dispersing water. Decon pools used during a hazardous materials incident work well as catchalls. These supplies should be carried along with the standard property-conservation equipment.

Waste Pipes

In some structures, mainly commercial and industrial buildings, waste pipes are exposed rather than covered behind walls. These waste pipes provide another effective means for removing large quantities of water; however, because the pipes must be broken to be used for water removal, small amounts of water should be removed by other means, and this method should be used only to remove large quantities of water from an area near the pipes.

The waste pipe is broken at floor level or as close to the floor as possible **Figure 10-12**. Any pieces of pipe that extend above the floor are broken off to make the opening even with the floor. Waste pipes in older structures are usually made of cast iron, whereas in newer structures, they can be cast iron, plastic, or a ceramic material. Any of these can be easily opened with forcible-entry tools. A drain screen should be placed over the opening to keep out debris. A suction strainer used on hard suction hose could possibly be used for this purpose.

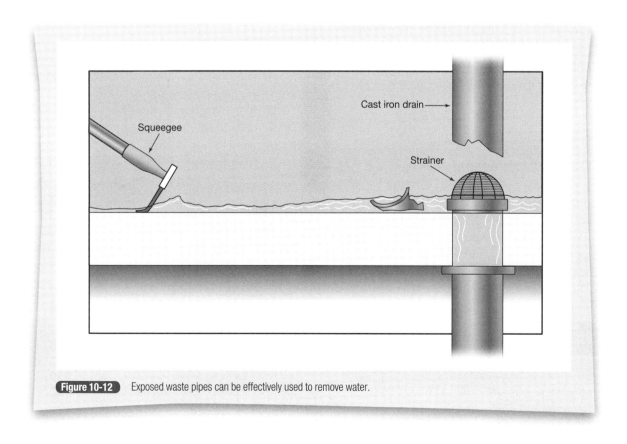

Figure 10-12 Exposed waste pipes can be effectively used to remove water.

Key Points

A building can usually be drained through the waste pipe on the fire floor, especially if it is opened soon after fire attack begins.

This operation can be repeated on several floors of the building, as necessary; however, the waste pipe on the uppermost floor must be opened first and that floor drained of water before the pipe is opened on a lower floor. If the waste pipes on two floors are opened at once, water from the upper floor will pour out onto the lower floor.

A building can usually be drained through the waste pipe on the fire floor, especially if it is opened soon after fire attack begins; however, there should be no hesitation in opening the waste pipes on lower floors if such action is required.

Openings in Walls

When large quantities of water are being used to extinguish a fire and water is accumulating quickly, large openings may be required to get the water out and keep the building safe. One such opening could be made by removing the outside wall immediately below a window from the sill to the floor. Hand tools, power tools, or both can be used, but ladder crews must be careful not to cut or damage structural members while removing the wall.

If possible, the windows below the opening should be closed or covered before the opening is made. Personnel below the opening must be warned to stay clear of falling debris and water.

This is an extreme action that should be used only under extreme conditions—when large amounts of water endanger the building and the fire fighters operating within it. In most instances, the wall cut will first be used to remove water from the floor below the fire while streams are still being directed onto the fire. Ladder crews engaged in such an operation must be kept informed of the fire situation above them and of the amount of water on the fire floor. They must also be on the lookout for signs of building collapse. Command must clearly communicate, by the best method, with fire fighters and safety officers regarding the water removal operation and the conditions within the building.

After engine companies have extinguished the fire near the windows on the fire floor, similar openings might have to be made

Key Points

Ladder crews must be careful not to cut or damage structural members while removing the wall.

Key Points

Pumps can be effective in removing water from areas of a building where other means are not feasible. Among these areas are basements, elevator or other shaft pits, and other low areas without drains.

there. With these two floors opened, the greatest accumulation of water can be removed.

Pumps

Various types and sizes of portable pumps, powered by electricity or gasoline, are available for removing water from buildings. They can be used alone or in combination with other water removal methods; however, most of these pumps have small capacities, and thus, they cannot be used for the quick removal of an appreciable amount of water.

Pumps can be effective in removing water from areas of a building where other means are not feasible. Among these areas are basements, elevator or other shaft pits, and other low areas without drains.

Elevator pits and large utility shaft pits can be used to quickly remove water from upper floors; however, this operation is effective only when the bottom of the pit is below the level of the basement floor. Water from above is directed down the shaft, and submersible pumps, portable pumps, or siphons are used to remove the water from the pit to the outside. Water can be removed from basements in the same way.

Elevator shafts should be used as a last resort to evacuate water from an upper floor under any type of water incident. Elevator operations must be shut down and not restarted until inspected by an agency authorized to do so.

Because of the great amount of debris in the water, pumpers should not be used in removing water from basements or pits. Venturi siphons will get the water out faster than drafting

Key Points

Elevator pits and large utility shaft pits can be used to quickly remove water from upper floors; however, this operation is effective only when the bottom of the pit is below the level of the basement floor.

Key Points

Elevator shafts should be used as a last resort to evacuate water from an upper floor under any type of water incident.

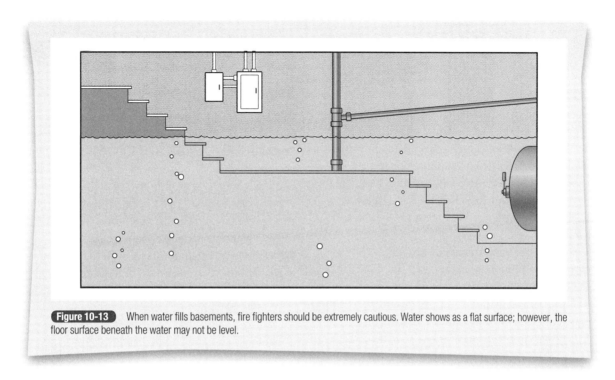

Figure 10-13 When water fills basements, fire fighters should be extremely cautious. Water shows as a flat surface; however, the floor surface beneath the water may not be level.

with a pump and without the danger of damaging fire apparatus.

Fire fighters should be extremely cautious when entering a water-filled basement. Although water always shows a flat surface, the floor beneath might not be level. For instance, there could be more than one level of basement flooring, as shown in **Figure 10-13**. Fire fighters must be very careful when entering a flooded area and should probe ahead with pike poles to determine the location of the floor, stairs, and other obstacles. Many fire fighters have been injured, and a few drowned, through a lack of caution when working in such areas. Such risks should not be taken that would endanger the life and safety of the firefighters.

Also, the water in a basement can get deep. Fire fighters should not put their faces under water or even close to the water surface to open a clogged drain. The suction that results when the drain is opened can pull under even the strongest fire fighter. Clogged drains must be cleared with the end of a tool.

Smoke and Property Conservation

During a fire incident, smoke can do considerable damage within the building and, depending on the situation, may exceed the damage caused by water. Initial ventilation operations conducted by fire fighters performing the primary search and fire extinguishment tactics will support property-conservation methods until a concerted effort is begun. Windows and doors can be opened for horizontal ventilation. Vertical ventilation can be accomplished by using existing roof openings, such as a penthouse, skylights, or scuttle hatches. Firefighters may have to open a roof to accomplish effective vertical ventilation. If this task is done primarily for property conservation, damage to the roof should be weighed against the smoke damage to the building and contents. PPV fans and blowers are used to remove smoke from the building. Efforts should be made to direct the smoke away from uninvolved areas of the building. Fans must not be allowed to enhance the fire or push smoke or fire toward escape routes or into unburned areas.

Key Points

Fire fighters may have to open a roof to accomplish effective vertical ventilation. If this task is done primarily for property conservation, damage to the roof should be weighed against the smoke damage to the building and contents.

Wrap-Up

Chief Concepts

- Property conservation is the third priority following life safety and extinguishment of the fire. It should be started as soon as possible after fire attack begins. If this operation is to be effective, ladder company personnel must understand why property-conservation efforts are initiated and, if assigned this task, know how to quickly, efficiently, and safely carry out the operation. In particular, they must know how to do the following:
  - Capture and contain water runoff
  - Channel water into drains or chutes and direct the water out of the building
  - Cover valuable property in the affected area or move it to another area or out of the building
  - Shut off the water supply to sprinkler systems when ordered to do so
  - Perform ventilation activities and operate PPV fans and blowers to remove products of combustion
- The safety of fire fighters engaged in all fire fighting activities, including property conservation, must be the number one priority. Command, as well as those firefighters involved in property-conservation efforts, must be cognizant of the fire conditions and of how it might affect their work.

Key Terms

Catchalls: Salvage covers rigged as basins to catch and hold water dripping from overhead, as from a ceiling.

Conduits: Salvage covers folded for use as conduits to direct accumulated water to stairways and then down the stairways and out through exterior doorways

Property conservation: Third tactical priority behind life safety and extinguishment. Fire fighters perform it to minimize the damage caused by water and smoke during a fire.

Salvage covers: Large sheets of waterproof material that are available in several materials, dimensions, and shapes. Some are fire resistant, and some are not.

1. Property conservation relates to overhaul operations in the following way:

   a. Property conservation is integrally related to overhaul operations, as they are generally performed at the same time and have the same purpose, namely reducing property loss.

   b. Property conservation is generally related to overhaul operations in that overhaul and salvage operations generally have the same purpose, but salvage (property conservation) should be started as soon as possible.

   c. The purpose of overhaul is to completely extinguish the fire, whereas the purpose of property conservation is to protect valuable property. The objectives are different, but they should be performed at the same time—that is, after all life safety and extinguishment efforts are complete.

   d. Property conservation and overhaul are not related; they have different objectives and are initiated at different times.

2. Property conservation is

   a. An important tactical priority that must be accomplished at all fires.

   b. An important tactic employed only during offensive operations

   c. An important tactic employed primarily during defensive operations

   d. An important tactic employed during offensive operations, but seldom used during defensive operations

3. Of the property-conservation methods listed here, which is the most labor intensive?

   a. Covering contents with salvage covers

   b. Removing water using water chutes

   c. Removing contents to the outside

   d. Capturing water using catchalls

4. Floor construction has a direct impact on water migration when dealing with a concrete floor:

   a. The primary protection against water damage involves covering contents on the floor below the fire.

   b. The primary protection against water damage involves containing the water and then removing it via chutes, floor drains, windows, or stairs.

   c. The primary property-conservation method is to limit the amount of water used to fight the fire.

   d. Actually, all three of the methods listed are of equal importance.

5. Catchalls are most effective in containing

   a. Large quantities of water from hose streams on the floor below the fire

   b. Large quantities of water at the scene of water leaks from internal plumbing

   c. Small quantities of water dripping from the overhead, as from a ceiling

   d. Water from sprinkler systems activation with a catchall placed below each head that is operating

6. The weight of one gallon of water is

   a. 0.434 kilograms

   b. 0.434 pounds

   c. 8.33 kilograms

   d. 8.33 pounds

7. The waste pipe opening from a toilet

   a. Should not be used to drain water from a floor area because the opening around the removed toilet will leak water to the floor below

   b. Should not be used to drain water from a floor area because the flooring materials used in bathrooms are very susceptible to water damage

   c. Can be used for very small quantities of water, but the trap in the pipe will slow water drainage

   d. Can be used to quickly remove much water from a floor area

8. In the past, some fire departments have made openings in outside walls to relieve water that is accumulating on the fire floor and floor below, this practice is

   a. Reserved for extreme situations when the quantity of water on the floors endanger building stability

   b. Limited to situations where a defensive operation is in progress

   c. A common practice and acceptable, as long as the opening in the outside wall does not affect structural stability

   d. An archaic method no longer suited to modern fire operations

9. Elevator shafts

   a. Are the best way to remove water from an upper floor provided the elevator car is located above the floor where water is being evacuated

   b. Are preferred when small quantities of water (not more than the capacity of the pit) are being removed and the elevator car is located above the floor where water is being evacuated

   c. Are better than using the stairs, but not as good as floor drains, including toilet waste pipes

   d. Are a last resort method of removing water

Elevated Streams

Chapter 11

Learning Objectives

- Recognize the procedure for providing an adequate water supply to an elevated master stream device.

- Evaluate nozzle selection and operating pressures when using an elevated master stream device.

- Consider stream placement during an offensive or defensive attack.

- Realize when it is appropriate to shut down an elevated master stream appliance.

- Understand situations in which an elevated master stream appliance should not be used.

- Consider the procedure of hose-line operations from an elevated aerial device.

Elevated streams can be used on the fire ground for a direct attack, **exposure protection,** or a combination of both. Precautions must be observed around buildings or other structures in which an elevated master stream is being directed. An example would be to use caution when operating near electrical power lines. Aerial apparatus should be strategically positioned so that the elevated stream can be used to its best advantage while providing a safe area for fire fighters to work in.

During defensive operations, aerial apparatus should be positioned so that radiant heat or structural collapse will not affect the safety of fire fighters working in the area. Elevated master streams should originate as far away from the fire as possible while still being effective. If there is any indication of structural integrity, a perimeter should be set up a safe distance from the collapse zone, and all personnel and equipment kept out of this area.

Command must manage the use of elevated master streams. Master stream appliances should not be placed in service in a defensive mode while an offensive, interior attack is being conducted. The use of elevated streams must be coordinated between all interior fire attack crews. Serious injury to fire fighters could result as fire, heat, and gases are forced down onto them from the effects of the stream. Elevated streams should never be directed into a building in which crews are operating in an offensive mode. In addition, elevated master stream appliances should not be operated into vertical, natural openings or openings created by fire fighters or the fire itself. This mistake will certainly endanger fire fighters inside the building and prevent the fire, heat, and gases from leaving the building.

Members operating hose lines from aerial ladders or elevating platforms must be attached to the aerial device with a fall protection harness. It is a dangerous task to take a hose line over an aerial ladder for operation from the device or to extend the hose line into the building. Although operations from an elevating platform may seem somewhat safer, a fall protection harness must be worn when operating or advancing a hose line into a building.

Fire fighters are subjected to dangerous conditions when operating a portable elevated master stream device (detachable ladder pipe) from the tip of an aerial ladder. No one should be positioned on the aerial ladder when the aerial device is being elevated, rotated, extended, or retracted. It is recommended that for aerial ladders without a permanent waterway attached the detachable ladder pipe be operated from the ground level using ropes or other remote controls attached to the nozzle.

During any operation involving elevated streams, fire fighters must work within the limitations of the aerial device following the prescribed manufacturer's recommendations for that particular aerial apparatus.

An important function of aerial fire apparatus is to provide an elevated stream for fire attack or exposure protection. The elevated stream is directed to the fire or exposure from a raised aerial device. In most cases, a master stream appliance will develop the stream. A hose line may also be advanced and operated from an aerial device if needed.

Elevated streams can be effectively directed into or onto the upper parts of tall buildings, beyond the reach of portable ground monitors or engine-mounted master stream appliances. They can also be placed in service on large, sprawling structures, outside storage yards such as lumber yards and oil storage tank farms, piers, ships, and, in general, areas where their height and reach provide access to the fire or exposure. To maximize this effectiveness, an aerial ladder or elevating platform's master stream appliance should be equipped with solid-bore nozzles of several diameters and at least one large-caliber spray nozzle. This will allow fire fighters to develop the proper stream for any combination of fire situation, weather conditions, and water supply.

Providing a Water Supply

Aerial ladder and elevating platform master stream appliances can, as noted, be used for direct fire attack, exposure protection, or a combination of both. If possible, the apparatus should be positioned with regard to both the wind direction and the location of the fire and exposures. A continuous, uninterrupted water supply must be provided to the master stream appliances. This water supply must be of sufficient volume to provide the master stream device appliance with the proper volume to perform the required task. Water supply is provided by engine companies or by aerial apparatus that are equipped with a fire pump and carry supply hose.

Spotting the Aerial Ladder

In some situations, a master stream appliance may have to be operated with the truck in an unfavorable position. For example, the aerial device might have been spotted for rescue or for ventilation operations while hose lines were being used for fire attack. Continued growth in the size and intensity of the fire could force command to change to a defensive mode of operations and place the master stream appliance in service. Because of the placement of pumpers and other fire apparatus on the fire ground, it may be impossible or impractical to reposition the truck. The master stream appliances would then be operated as effectively as possible from its original position. If it is obvious that the magnitude of the fire could place fire fighters

Key Points

Aerial ladder and elevating platform master stream appliances can be used for direct fire attack, exposure protection, or a combination of both.

Key Points

Where wind is not a factor, the aerial device should be spotted for maximum coverage of the fire area. The location of the fire and its direction of travel in the building are critical factors in determining the best position for the truck.

and apparatus in harm's way, then all fire fighters and apparatus should be relocated expeditiously to a safe position.

In some situations, however, the operation of the master stream appliance will begin with arrival on the fire ground. If this is the case, the aerial device should be spotted for maximum effectiveness.

Buildings　Where wind is not a factor, the aerial device should be spotted for maximum coverage of the fire area, as shown in **Figure 11-1**. The location of the fire and its direction of travel in

the building are critical factors in determining the best position for the truck.

The **aerial device** should be spotted to operate between the fire and the exposures when the wind is blowing across the face of the fire building so that nearby structures are exposed (see **Figure 11-2**). This will allow the elevated stream to be directed to the fire building and, at the same time, to be in position to cover the exposures in case there is danger of further fire extension.

Open Storage Areas　Here, again, when wind is not a factor, the aerial device should be spotted depending on the location of the fire and the direction of travel. Exposure protection should also be taken into consideration.

When there is a wind, especially a strong one, however, the aerial device should be spotted at the flank of the fire between the main body of fire and exposures to attack the fire from the unburned side, as shown in **Figure 11-3**. If elevated master streams are properly positioned, they can be directed onto the

Figure 11-1　The aerial device should be spotted or placed for maximum coverage of the fire area if wind is not a factor.

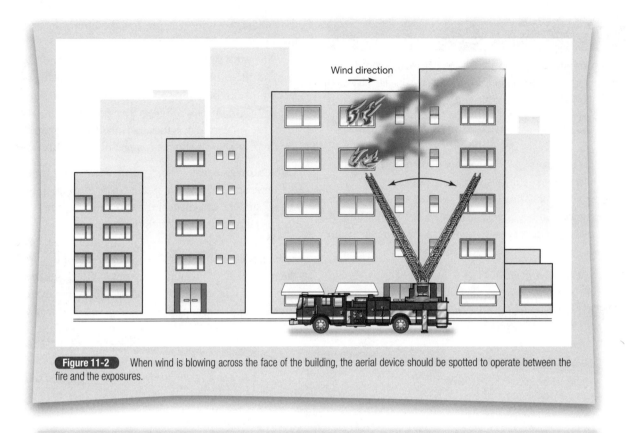

Figure 11-2   When wind is blowing across the face of the building, the aerial device should be spotted to operate between the fire and the exposures.

Figure 11-3   When operating at large open fires, aerial apparatus should be placed at the flanks of the fire, if possible, to operate elevated master streams to protect exposures and extinguish the fire.

Key Points

When there is a wind, especially a strong one, the aerial device should be spotted at the flank of the fire between the main body of fire and exposures to attack the fire from the unburned side.

fire and the exposures, and water spray caught by the wind will fall on the exposures. At the same time, the aerial apparatus will not be endangered if the elevated streams cannot control the fire. Aerial apparatus should not be positioned directly in the path of the fire. If the apparatus was positioned directly in the line of fire travel and the elevated streams were unable to control the fire, both fire fighters and apparatus would be left in a precarious position.

Elevated streams also could be used to protect fire fighters advancing hose lines toward a fire in an open storage area. This operation especially can be important when the radiant heat from the main body of fire is intense or when the hose lines must be advanced around piles of stored material. The aerial device should be spotted either behind or at the flanks of the advancing fire fighters. The aerial ladder or elevating platform should be raised over the crew and streams directed just ahead of them as they advance.

Flammable Liquid–Handling Facilities   Elevated streams can be effective at fires involving flammable liquids and gases. The storage area positioning discussed previously should be used at facilities where flammable liquids and gases are processed or stored; however, the aerial device should never be spotted in line with either end of a horizontal tank or cylinder. This is an extremely dangerous position, as the ends of the tank or cylinder could blow out if the container explodes. If possible, the aerial apparatus should be positioned at the side of the tank or cylinder, and the elevated stream used to cool the container as well as to attack the main body of fire, as shown in **Figure 11-4** .

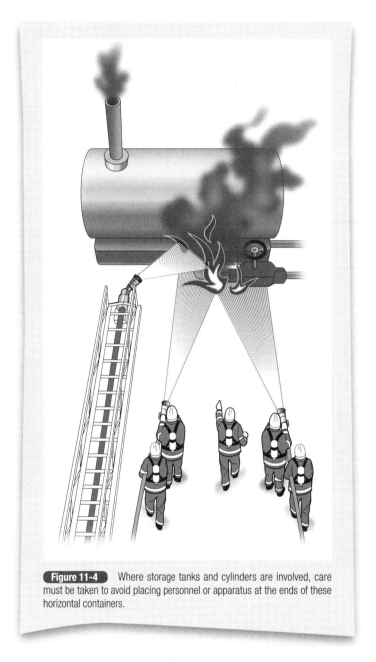

Figure 11-4   Where storage tanks and cylinders are involved, care must be taken to avoid placing personnel or apparatus at the ends of these horizontal containers.

Key Points

Elevated streams also could be used to protect fire fighters advancing hose lines toward a fire in an open storage area.

Key Points

The aerial device should never be spotted in line with either end of a horizontal tank or cylinder. This is an extremely dangerous position, as the ends of the tank or cylinder could blow out if the container explodes.

Water Supply

The water supply for aerial master stream appliances is most often developed in cooperation with engine companies. Engine crews usually obtain the water from a hydrant system, drafting sites, or possibly a fireboat. The engine company lays the supply hose to the aerial apparatus and provides the pumping pressure.

Key Points

The water supply for aerial master stream appliances is most often developed in cooperation with engine companies. Engine crews usually obtain the water from a hydrant system, drafting sites, or possibly a fireboat.

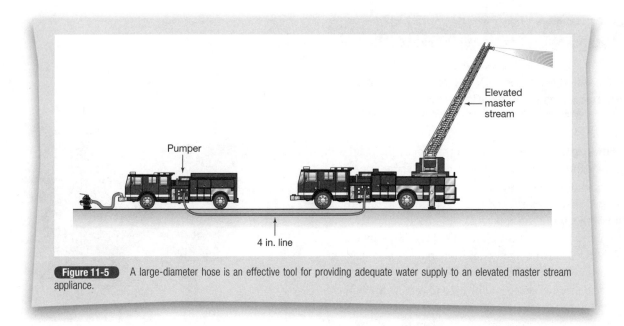

Figure 11-5 A large-diameter hose is an effective tool for providing adequate water supply to an elevated master stream appliance.

Ladder crews then are responsible for moving the water from the intake connections on the truck through the waterway to the master stream appliance itself.

The use of large-diameter hose to supply an elevated master stream appliance typically has replaced the use of several smaller supply lines to achieve the same results **Figure 11-5**. Elevated master stream appliances have the capability of flowing large amounts of water. A 2-inch solid-bore nozzle at 80-psi nozzle pressure will flow 1,000 gallons of water. To obtain this flow, supply lines must be capable of providing a volume of water sufficient to conduct the operation.

The water supply operation will be effective if engine and ladder crews are able to work well together. As in most firefighting operations, a coordinated effort is required between engine and ladder companies. Operating a master stream appliance from an aerial device should be practiced on the training ground, where fireground problems can be duplicated and where time is available to evaluate and solve these problems. Mistakes made on the training ground can be corrected so that they do not occur on the fire ground. Moreover, as training progresses and skills improve, engine companies will become aware of the problems facing ladder companies and vice versa. When each group is able to anticipate the needs and movements of the other and operate accordingly, the net effect will be increased efficiency in all joint fireground operations.

Water Delivery Systems for Aerial Ladders

Most modern aerial apparatus are equipped with a permanently mounted water delivery system. This system eliminates the task of operating an elevated master stream appliance for a detachable ladder pipe system during initial attack operations.

Rigging a Pipe System   Where a prepiped waterway is not provided, the following procedure can be used. A ladder pipe needs to be attached to the aerial ladder. Sufficient lengths of 3-inch hose or larger diameter hose, complying with the requirements of NFPA 1961, *Standard on Fire Hose,* are used to reach between the installed ladder pipe and the ground. If 2½- or 3-inch hose is being used, a valved three- or four-inlet siamese should be provided for water supply. A hose strap should be used on each section of the aerial ladder to secure the hose line. Halyards are connected to the ladder pipe to control the operation from the ground. It takes considerable time and personnel to set up this operation on the fire ground.

Prepiped Systems   A permanently mounted water delivery system allows aerial apparatus to get a master stream into operation quickly and with little effort. There are two basic prepiped waterway systems.

A **telescoping waterway** extends up from the base of the bed ladder to the master stream appliance located at the tip of the fly section. This allows the master stream appliance to be operated

Key Points

The use of large diameter hose to supply an elevated master stream appliance has typically replaced the use of several smaller supply lines to achieve the same results.

Key Points

A permanently mounted water delivery system allows aerial apparatus to get a master stream in operation quickly and with little effort.

from any height from the tip of the bed section to the tip of the fully extended fly section. The waterway's piping reduces in size as each section extends as the aerial is extended. At the base of the aerial, the waterway is connected to additional piping containing a swivel. The swivel joint allows the aerial ladder to be rotated 360 degrees. The piping continues through the turntable assembly to external suction inlets, which are usually located at the rear or sides of the apparatus. If the ladder truck is equipped with a fire pump capable of supplying the required flow and pressure, a permanent valve connection shall be provided between the pump and the waterway system.

In a nontelescoping system, the waterway extends from the base of the bed ladder to the master stream appliance located at the tip of the bed ladder. This system allows a master stream appliance to be operated from only the tip of the bed section. The use of this system obviously can limit the reach and effectiveness during the operation. Water is supplied to the master stream appliance in the same manner as for the telescoping waterway.

Water Delivery Systems for Elevating Platforms

The waterway system for an aerial apparatus containing an elevating platform is similar to that of a conventional aerial ladder. The master stream appliance(s) on an elevating platform is supplied by a permanent water system. The design of the modern aerial tower or elevating platform waterway is telescoping. The difference is that the master stream appliance is located in the platform, whereas the master stream appliance on an aerial ladder is located at the end, or tip, of the ladder. An elevating platform may contain an additional master stream appliance that would allow both to be operated at the same time if necessary.

For safety, the waterway should be charged after the tower is extended and before the platform is moved toward the fire. The master stream and the water curtain system under the basket then will be available to protect fire fighters working in the platform as well as the apparatus if fire should travel toward the unit **Figure 11-6**. The water curtain system is designed to provide a cooling spray under the entire floor of the platform. Provisions should be made so that personnel working in the platform can attach themselves to fall protection harnesses.

Elevated Streams for Fire Attack

To knock down and extinguish a fire, elevated streams must reach the seat of the fire. Because the aerial apparatus is not designed to be operated inside a building, the master stream appliance must

Key Points

For safety, the waterway should be charged before the platform is moved toward the fire.

Figure 11-6 Before the platform is moved in toward the fire, the waterway should be charged.

be operated from outside of the building. To ensure that a master stream appliance is used to its best advantage, ladder crews must ensure that the properly sized nozzle is used to discharge an adequate amount of water at a proper operating pressure.

Nozzles

As noted earlier, both spray nozzles and solid-bore nozzles should be carried on the apparatus for use with the master stream

Key Points

To ensure that a master stream appliance is used to its best advantage, ladder crews must ensure that the proper size nozzle is used to discharge an adequate amount of water at a proper operating pressure.

appliance. Depending on the desired application, either nozzle could be used. Solid-bore nozzles provide an excellent stream to penetrate into the building and reach the seat of the fire. This can be beneficial for buildings of significant size and area. Because of the reach of the stream, the aerial apparatus can be positioned at a greater distance from the building. This tactic may be used if the fire is large and conditions are not conducive to positioning apparatus close to the fire. Under normal conditions, solid streams are not affected by wind conditions and do not break up before reaching the fire. They are less affected by steam conversion than spray nozzles. Spray nozzles, on the other hand, have the ability to operate from a straight stream to wide-angle fog or water spray. Spray nozzles can be affected by wind conditions, which may break up the stream, causing it to be less effective. Fog patterns from spray nozzles do not have the penetrating power to provide the reach or penetration as solid-bore nozzles. This may require the apparatus to be positioned closer to the fire building. The correct type of nozzle should be selected based on the best tactics needed for the fire-control situation.

Both spray nozzles and solid-bore nozzles are rated according to their water flow rates in gallons per minute. Either type will lose effectiveness if it is not supplied with at least its minimum rated flow. For example, a 500-gpm nozzle supplied with 500 gpm will be more effective than a 1,000-gpm nozzle not receiving a sufficient water supply. For this reason, the master stream appliance should be equipped with the proper size as well as the proper type of nozzle necessary for that particular operation. This is why a range of sizes of spray fog and solid-stream nozzles should be carried on the apparatus.

To ensure that the master stream appliance can achieve its rated capacity, the water delivery system must be capable of delivering 1,000 gpm at 100 psi nozzle pressure with the aerial ladder or elevating platform at its rated vertical height. As noted earlier, the aerial apparatus must be furnished with a continuous, uninterrupted water supply of adequate volume and pressure to maintain the operation.

Key Points

Both spray nozzles and solid-bore nozzles are rated according to their water flow rates in gallons per minute. Either type will lose effectiveness if it is not supplied with at least its rated flow.

Key Points

To ensure that the master stream appliance can achieve its rated capacity, the water delivery system must be capable of delivering 1,000 gpm at 100 psi nozzle pressure with the aerial ladder or elevating platform at its rated vertical height.

Stream Placement

To direct a spray nozzle's stream into a fire building through a window, place the stream at the approximate center of the window opening and set the nozzle at a 30-degree angle. Aim it first at the upper part of the room, where the concentration of heat is greatest, and then sweep downward. Repeat this action as necessary to control the fire.

Place a solid-bore nozzle's stream so that it enters the window at an upward angle. This will allow the stream to strike the ceiling, break up, and spread water over a wide area **Figure 11-7**. For maximum penetration, theoretically the stream should enter just over the window sill at a fairly small angle; however, the stream might then hit the building contents rather than the ceiling. A somewhat larger entry angle provides good penetration and will probably allow the stream to clear interior obstacles. Too large an entry angle will lose penetration, and the water will cascade down just inside the window.

Direct a solid stream straight into a window when maximum penetration is the most important consideration; however, never direct it down toward the floor from a position above the window; this will have no effect on the fire and will only add weight to a structure probably already weakened by the fire. A water flow rate of 500 gpm adds about 2 tons of water to the building in 1 minute.

To be most effective, the elevated stream, either spray or solid stream, should be moved horizontally back and forth across the fire area. Also, move it up and down for maximum coverage. The amount of movement depends on the extent of the fire, but normally, solid streams should be moved across the full width

Key Points

Place a solid-bore nozzle's stream so that it enters the window at an upward angle. This will allow the stream to strike the ceiling, break up, and spread water over a wide area.

Key Points

Direct a solid stream straight into a window when maximum penetration is the most important consideration.

Key Points

To be most effective, the elevated stream, either spray or solid stream, should be moved horizontally back and forth across the fire area. Also move it up and down, for maximum coverage.

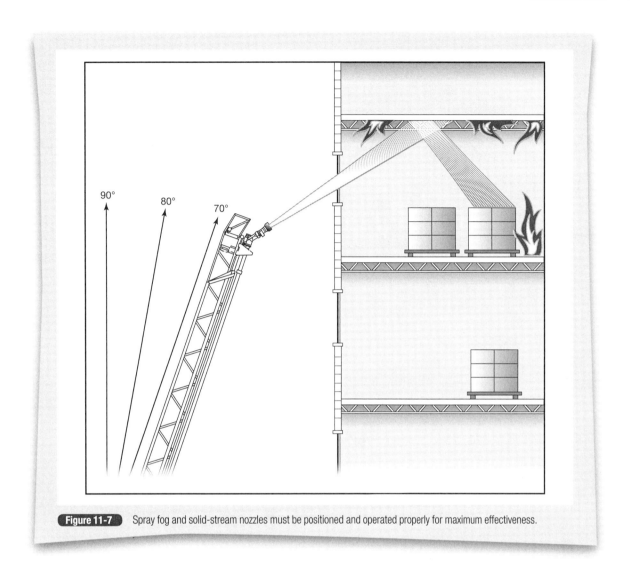

Figure 11-7 Spray fog and solid-stream nozzles must be positioned and operated properly for maximum effectiveness.

and height of the window. Although nothing can burn directly under a stationary stream, the fire can spread away from it, so movement of the stream ensures coverage of a good-sized area in and around the fire.

In heavy smoke, it may be difficult to determine whether the stream is entering the building. Fire fighters should look for steam and white smoke as indications that the stream is penetrating the fire area. In the absence of these signs and if it is possible to do so safely, an officer or fire fighter should visually check the target area of the stream. This can be accomplished by standing off to the side and observing the direction of the stream. It should not require personnel to get close enough to a building or other hazard where they may be endangered. If a visual check is not feasible, they should listen for the sound of the stream hitting the building and should look for heavy water runoff. Either of these signs indicates that the stream is not entering the building. The stream should then be adjusted until both indications have disappeared.

Wind and Thermal-Updraft Effects

Sometimes the master stream operated from aerial apparatus is adversely affected by the thermal updraft created by a large free-burning fire, as shown in **Figure 11-8a**. It might also be affected by wind blowing across the stream, as shown in **Figure 11-8b**, or toward the apparatus. Either the updraft or the wind will tend to break up the stream, reduce its reach, and thus render it ineffective. Placing the nozzle close to the window or using a straight stream pattern, as shown in **Figure 11-8c**, will alleviate the problem.

Fog patterns, created by spray nozzles, are made up of smaller and lighter water droplets and are subject to wind and thermal updrafts to a greater degree than solid streams. The wider the fog pattern, the more it will be affected. If a fog stream is being broken up by winds or thermal updrafts, the nozzle can be adjusted to a narrower pattern, 30 degrees being the most effective, or to the straight stream position; however, a straight stream from a spray nozzle is not as effective as one from a solid-bore nozzle.

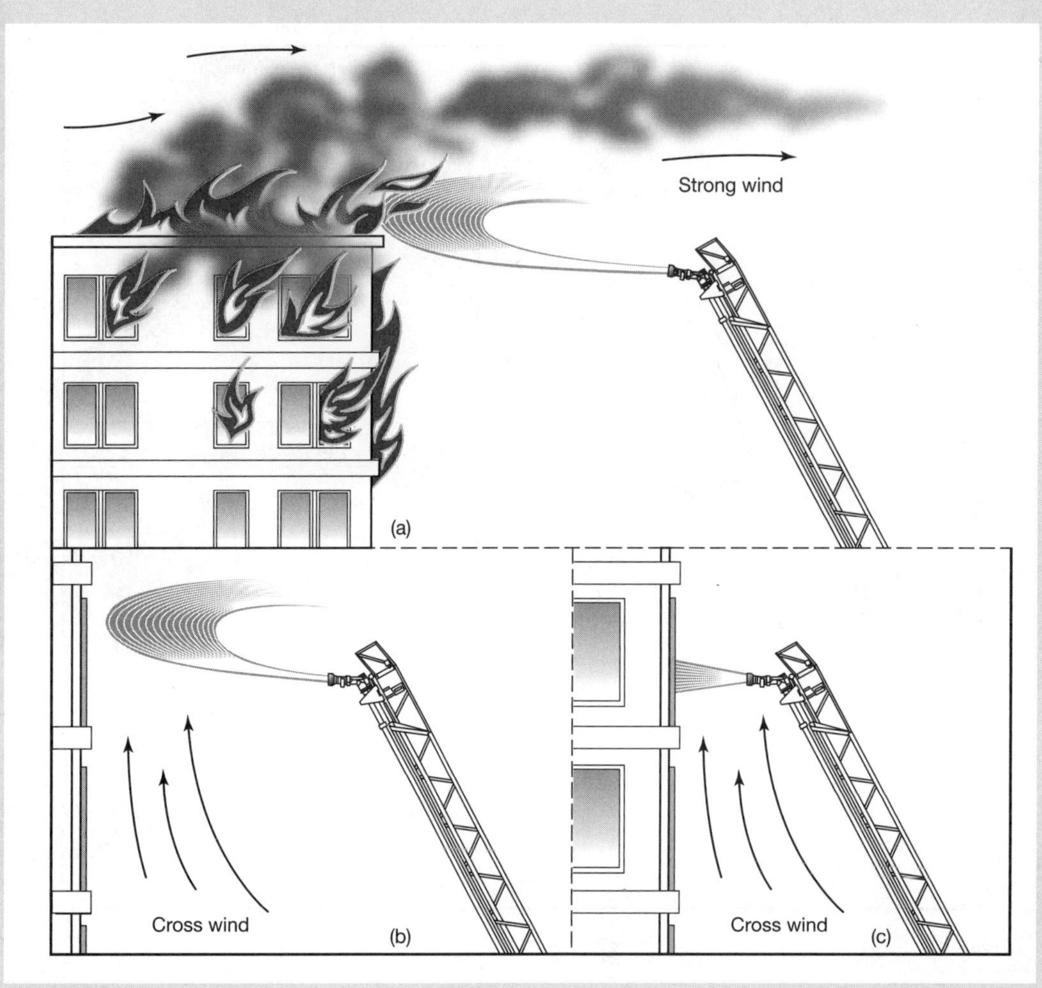

Figure 11-8 Strong winds will adversely affect elevated fog pattern streams, but placing the nozzle close to the window or using a straight stream pattern will alleviate the problem.

Move the spray nozzle very close to the fire building, inside a window if possible. If the fire situation prevents this, shut down the appliance, and replace the spray nozzle with as large a solid-bore nozzle that can be adequately supplied with water. The solid stream will hold together much better than the fog stream, and a heavier stream will hold better than a lighter one.

Weakened Structures

At most structural fires where elevated streams must be used, the streams will be directed into the structure from positions close to it. If the structure shows signs of having been weakened by

the fire or if chimneys or other heavy items on the roof such as HVAC units or roof-mounted billboards or other features seem ready to collapse, the aerial apparatus must be moved away from the building. This may place the master stream appliance in a less-desirable position but is necessary to protect the equipment and fire fighters.

When a spray nozzle is being used to fight the fire, check it after the aerial apparatus is moved to determine whether the stream is still effective. If the stream cannot reach into the fire

Key Points

If a fog stream is being broken up by winds or thermal updrafts, the nozzle can be adjusted to a narrower pattern, 30 degrees being the most effective, or to the straight stream position.

Key Points

If the structure shows signs of having been weakened by the fire or if chimneys, roof-mounted billboards, or other features seem ready to collapse, the aerial apparatus must be moved away from the building.

from the new position, replace the spray nozzle with a solid-bore nozzle, which should also be checked. If the solid stream is helping control the fire, operate it until the fire is knocked down. If not, the master stream appliance should be shut down. It obviously is ineffective and the water that is being wasted in this operation may be better put to use elsewhere on the fire ground.

Shutdown

Elevated streams should be used only as long as fire, steam, or white smoke is visible in the area covered by the stream. The steam and white smoke indicate that the stream is hitting the fire. When they are no longer visible, the fire has apparently been put out in that area, and the streams should be shut down. Continued operation would only add to the water load in the building and the strain on the water supply system.

Key Points

Elevated streams should be used only as long as fire, steam, or white smoke is visible in the area covered by the stream.

Improper Use of Streams

An incorrectly used elevated stream can cause unnecessary property loss and can result in injury to fire fighters. The most common error is directing elevated streams through roof openings, whether they are natural openings or created by fire fighters or the fire. This error is compounded significantly when fire fighters are operating in the offensive mode inside the building. There can be no excuse for combining offensive and defensive modes of operation on the fire ground.

Roof Openings Elevated streams should not be directed into an opening that has been created for ventilation or has been burned through by the fire. The streams will destroy the venting action and drive heat, smoke, and gases back into and through the building **Figure 11-9**. Even when fire shows through the roof along with vented combustion products, elevated streams should

Key Points

An incorrectly used elevated stream can cause unnecessary property loss and can result in injury to fire fighters.

A stream pushing everything back into the building and spreading the fire condition.

Figure 11-9 Elevated streams must not be directed through roof openings whether they are natural openings or created by the fire fighters or the fire.

not be directed into the roof hole. This process is supposed to occur. The fire, heat, and products of combustion rise and leave the building through these openings. It is counterproductive to direct a stream of water into these openings and push everything that is trying to leave the building back in.

To protect a roof from ignition, direct an elevated stream onto the roof near the opening. This will allow the water to

flow around the opening and down along the edges of the hole. The outside of the roof will be protected, and the combustion products still will be able to escape.

When a roof or a good portion of it collapses, it may be that only elevated streams are able to control the fire in the area of the collapse. An elevated stream can be used in such a situation because, usually, the roof opening will be large enough so that the stream does not interfere with the venting action.

Offensive Operations Inside of a Structure   Elevated streams should never be directed into a building area in which crews are operating in an offensive mode. Elevated streams may endanger fire fighters inside a structure by pushing fire, heat, and the products of combustion back into the building **Figure 11-10**. If the elevated streams are effective, great volumes of steam could engulf those inside the building.

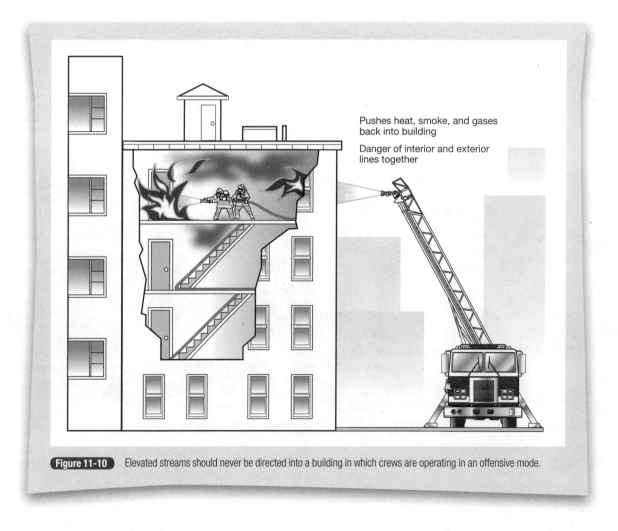

Pushes heat, smoke, and gases back into building

Danger of interior and exterior lines together

Figure 11-10   Elevated streams should never be directed into a building in which crews are operating in an offensive mode.

If command has chosen to operate in the defensive mode on arrival, an elevated stream may be used in an attempt to knock down the fire. During this operation, no one is allowed to enter the structure for any type of offensive operation. Hose lines may be stretched to exposures or in preparation to moving from a defensive mode to an offensive mode of operation.

In most cases, hose lines are advanced to support the defensive attack. They may be positioned in locations from which they can hit fire inaccessible by the elevated stream owing, perhaps, to the depth of the building or interior construction features. Such a combined attack, conducted in the defensive mode, could be most effective in knocking down the main body of fire. After the fire has been knocked down or greatly reduced, any operating elevated stream, or other stream operated from the outside, must be shut down before going from a defensive mode to an offensive mode of operation. All personnel on the fire ground must be notified via the chain of command when changing modes.

Elevated Streams for Exposure Coverage

Exposure protection is second only to rescue as a basic firefighting objective. The importance of the ladder company in checking for fire extension and in protecting interior exposures was discussed in Chapter 6. The protection of exterior exposures, including structures or other objects separate from the fire building, is discussed here. Objects refer to any number of items, including motor vehicles, tanks and cylinders, construction material, and grassland or forest areas.

Outside Exposures

The term **exposure fire** is applied here to outside exposures. Such a fire spreads from one structure to another or from one independent part of a building to another part, perhaps across a courtyard or from one building wing to another.

Spaces between buildings or other objects and unpierced fire walls are the major deterrents to exposure fires. They are of great assistance to fire fighters when severe outside fires develop. Outside sprinklers and spray systems are also a great help, but unfortunately, they are a rare item in fire protection equipment except in special installations.

Exposure Hazards

Ladder company officers and crews, as well as engine company personnel, must be familiar with potential exposure problems, conditions that promote the exterior spread of fire, within their response area. They should also be cognizant of the factors that affect the severity of an outside exposure problem. These factors include the following:

- Recent weather
- Present weather, especially wind conditions
- Spacing between the fire and the exposures
- Building construction materials and design
- Intensity and size of the fire
- Location of the fire
- Availability and combustibility of fuel
- Size of the initial fire force and additional resources
- Firefighting equipment on hand

The worst combination would be recent dry weather, strong winds blowing toward exposures, an area of closely spaced frame buildings, a severe fire that is difficult to reach, plenty of easily ignited materials located between the fire building and exposures, limited personnel and apparatus response to the first alarm, insufficient access to special need apparatus, and a poor water supply.

Of these factors, the fire department normally has control of only the responding fire force and the equipment responding to the first and additional alarms. For this reason, the first-alarm assignments of ladder companies as well as the other companies should be reviewed periodically, keeping in mind the contribution that these units can make toward exposure protection.

Where the exposure hazards are great, the number of companies responding to a first alarm should be increased. It is far better for these units to be on the scene when they are not needed than to have them in the station when they are needed on the fire ground. This is especially true when ladder companies are located some distance from each other, as is usually the case with mutual-aid companies.

Key Points

Where the exposure hazards are great, the number of companies responding to a first alarm should be increased.

Exposure Protection

Exposure fires can be caused by radiated heat, convected heat, or a combination of both. Radiated heat is the transfer of heat through space in the form or energy waves. Convected heat is the transfer of heat through the mixing or circulation with air.

Properly positioned and operated elevated streams can do much to protect against exterior fire extension. Where wind is a factor, exposures on the downwind side of the fire are in the most danger from radiated heat and from embers carried by convection and wind currents. When a choice must be made, downwind exposures should be protected first, before exposures on the windward side are covered. During a severe fire, the fire itself can create a strong wind.

Key Points

Where wind is a factor, exposures on the downwind side of the fire are in the most danger from radiated heat and from embers carried by convection and wind currents.

Choosing the Stream   Fog streams are more effective than solid streams for exposure protection, provided that they can reach the area to be protected. If they cannot because of the distance to the exposure or because of wind or thermal-updraft effects, solid streams should be used.

Key Points

Fog streams are more effective than solid streams for exposure protection, provided that they can reach the area to be protected.

Embers

Convected heat

BOWLING

Direct stream against walls and windows of exposed building

Figure 11-11   Elevated streams must be directed on the area to be protected, not thrown between the fire and the exposure.

In general, the greater the intensity of the fire, the heavier the elevated exposure stream needed; however, no stream will be effective with a less-than-adequate water supply. If water supply is a problem, a smaller, adequately supplied stream will be more effective than a weak stream from a larger nozzle.

Directing the Stream   Because water is transparent, radiant heat passes through it. Simply throwing a stream between the fire and the exposure will not protect the exposure; the heat will simply pass through the stream and heat the surface of the exposure to its ignition point. Instead, the stream must be directed onto the surface of the exposure in such a way that the water washes down its sides **Figure 11-11**. Only then will the water absorb heat from the exposure and keep it from igniting.

If an exposed building is taller than the fire building, its most vulnerable area is above the level of the fire. That area will be subject to both radiated and convected heat and also to any firebrands and embers carried by the wind **Figure 11-12**. The first elevated stream in service should be directed onto the exposure just above this most vulnerable area.

Burning embers and ignited materials can be convected up to exposures from a fire that has burned through the roof of a building or has ignited other objects. A properly positioned elevated stream, directed into the fire and its smoke column, can be of great help in decreasing this exposure hazard. An exposure stream can often be used for this purpose; that is, the stream can be alternately directed onto the exposure and the fire. In many other situations, such action can help knock down the fire or decrease its intensity in the area of the exposure; however, the operator must be careful not to give too much attention to the fire and too little to the exposure.

When an exposure is so large that one elevated stream cannot protect it completely, an attempt must be made to position additional streams for complete coverage. If an elevated

Figure 11-12 The most vulnerable area on the exposure building is the area above the level of the fire.

Key Points

When an exposure is so long that one elevated stream cannot protect it completely, an attempt must be made to position additional streams for complete coverage.

stream cannot be positioned to cover the unprotected part of the exposure, other master stream appliances or 2½-inch hose lines must be used. In many cases, a combination of elevated streams, other master stream appliances, and hose lines will be required for complete exposure coverage and fire control.

Elevated Hose Lines

An elevated master stream appliance is a device that can deliver large volumes of water. In some situations, an elevated stream, but not a master stream, is required. Hose lines can be operated from an aerial ladder or elevating platform in these cases.

Operating from an Elevating Platform

Elevating platforms are usually provided with separate outlets to operate hose lines. These outlets may have 2½-inch discharges that reduce down to 1½ inch. In this way, either 2½-inch hose or smaller diameter 1¾-inch or 1½-inch hose can be operated from the platform. If a hose line is to be operated from the platform itself, care should be taken by fire fighters in maneuvering the hose line in the platform. Fire fighters should be attached to the platform with fall protection harnesses. A smaller length of hose with nozzle attached should be considered if working directly from the platform. In this way, excess hose will be kept to a minimum within the platform. The nozzle and about 1 foot of hose should extend out in front of the platform railing; this will allow the stream to be moved in all directions.

If the stream is to be directed in through a window, place the platform so the nozzle is at the center slightly below the

Key Points

In some situations, an elevated stream, but not a master stream, is required. Hose lines can be operated from an aerial ladder or elevating platform in these cases.

Key Points

The nozzle and about 1 foot of hose should extend out in front of the platform railing; this will allow the stream to be moved in all directions.

bottom part of the window. A fog stream at a 30-degree angle can then be aimed into the upper part of the room. A solid stream can be aimed for maximum penetration, and the operator will be safe from fire or gases that might be pushed out the window.

After the fire in the area of the window is knocked down, the platform can be moved above the sill to allow horizontal penetration, if necessary. The hose line may be used to wet down areas on the outside of the building as well.

Hose lines may also be advanced into a building from discharge outlets located on the platform. The platform is raised to a convenient entrance point where the hose line is advanced from the discharge outlet on the elevating platform into the structure. The hose line should be stretched from the platform outlet and when in position the line should be charged.

A department could have hose on the platform preconnected to the discharge outlet, saving time when this operation is to be used. The elevating platform can be used as a standpipe for a number of applications such as an occupied building, including the roof, a building under construction, a parking garage, and overhead structures such as a bridge, elevated roadway, or railway.

Operating from an Aerial Ladder

A hose line may be taken over an aerial ladder and advanced into a building or other structure for firefighting purposes. The aerial ladder should be positioned at the point where the hose line will be advanced. A hose line is advanced over the ladder from the ground to the point of entry. Enough hose needs to be advanced to the area of operation. The hose should then be secured to the ladder using hose straps, which take the strain off couplings and fittings when charged. Fire fighters working on the ladder should be provided with fall protection harnesses.

The ladder must not be extended or retracted once a fire fighter is on it. The ladder operator must be careful not to activate the extension–retraction control when rotating or moving the ladder, as this could seriously injure any fire fighters on the ladder.

During any operation involving elevated streams, fire fighters must work within the limitations of the aerial apparatus following the prescribed manufacturer's recommendations for that aerial

Key Points

A hose line may be taken over an aerial ladder and advanced into a building or other structure for firefighting purposes. The aerial ladder should be positioned at the point where the hose line will be advanced.

device. These procedures include the manufacturer's operating ranges for angle, extension and elevation, loading requirements, and flow ratings. This is an important factor during aerial operations. It is not difficult to exceed the manufacturer's limits and specifications. The aerial operator must be knowledgeable and aware of these limits at all times.

Key Points

During any operation involving elevated streams, fire fighters must work within the limitations of the aerial device following the prescribed manufacturer's recommendations for that aerial apparatus.

Wrap-Up

Chief Concepts

- Elevated streams provided by aerial apparatus are important in attacking a fire and protecting exposures. Ladder crews must know how to quickly set up their master stream appliances and coordinate with engines companies to provide a continuous, uninterrupted supply of water to the device. Officers and aerial apparatus operators must be able to accurately size up any given fire situation and spot the aerial device properly for fire attack, exposure protection, or both. They must know the difference between an offensive mode and a defensive mode of operation, which will dictate as to how an elevated stream shall or shall not be used.

- A nozzle of the proper size and type must be chosen on the basis of the intended use of the stream, required reach, water supply, wind conditions, and the effect the fire could have on the stream. Aerial apparatus should be able to place an elevated stream into operation using either a spray nozzle or a solid-bore nozzle. In most cases, solid streams of the proper caliber will be most effective.

- Elevated hose lines can be operated from aerial platforms using discharges supplied in the platform. Hose lines may also be advanced from an elevating platform and over aerial ladders into a building or other structure where a temporary standpipe is needed.

- Fire fighters must be familiar with and follow the manufacturer's written procedures when operating aerial apparatus during any operation. These procedures include the manufacturer's operating ranges for angle, extension, and elevation, loading requirements, and flow ratings.

Key Terms

Aerial device: Any ladder, platform, pumper, or other apparatus designed and operated to support water, rescue, or fire fighter operations.

Exposure fire: A fire that spreads from one structure to another or from one independent part of a building to another part from the original fire.

Exposure protection: The protection of other exposed structures by water streams.

Telescoping waterway: Extending from the base of the bed ladder to the master stream appliance located at the tip of the fly section; allows the master stream appliance to be operated from any height from the tip of the bed section to the tip of the fully extended fly section.

1. When the wind is blowing across the face of the fire building so that nearby buildings are exposed, the apparatus aerial device should be spotted:
   a. Upwind just beyond the base of the flames
   b. Downwind well beyond the flame front
   c. In front of the exposure building toward the fire building side
   d. Between the fire and exposure buildings

2. When confronted with a fire in an open area with high wind conditions, elevated master streams
   a. Should not be used
   b. Can be used, but should be placed at the flanks of the fire
   c. Can be used if placed on the downwind side well ahead of the flame front
   d. Can be used if placed on the upwind side

3. An elevated stream is being used to cool a fire impinging on a horizontal tank containing flammable liquids. The aerial device should not be placed at the _____ of a horizontal tank.
   a. Ends
   b. Sides
   c. Windward side
   d. Leeward side

4. A 2-inch solid-bore nozzle will flow _____ gpm at 80 psi nozzle pressure.
   a. 500
   b. 750
   c. 1,000
   d. 1,250

5. When using a spray nozzle to attack a fire inside a structure, place the nozzle at the approximate center of the window on a _____ pattern.
   a. Straight-stream
   b. 30-degree angle
   c. 60-degree angle
   d. Full fog

6. If the fog pattern on a spray nozzle is being adversely affected by wind, a _____ pattern should be used and the nozzle moved closer to the window.
   a. 60-degree or smaller angle
   b. 30-degree or straight stream
   c. Straight stream
   d. When wind is a significant factor, always discontinue the use of the spray nozzle and attach the solid-bore nozzle.

7. Elevated streams should continue to operate until _____ is no longer visible.
   a. Fire
   b. White smoke
   c. Steam
   d. The stream should continue to operate as long as any of the above are visible.

8. When fire first breaks through the roof and is visible at the roof opening
   a. Operate a solid-stream nozzle directly into the opening
   b. Operate a spray nozzle on full vapor pattern directly into the opening
   c. Operate a spray nozzle on straight stream pattern directly into the opening
   d. Do not operate an elevated stream into the roof opening

9. Elevated master streams should be directed into a building during
   a. Defensive operations only
   b. Offensive operations only
   c. Offensive operations, but not into areas where fire fighters are working
   d. Combination offensive/defensive operations

10. Fog streams are _____ effective than/as solid streams for exposure protection.
    a. Equally
    b. Less
    c. More
    d. Fog streams are more effective on frame construction, and solid streams are more effective on masonry construction.

11. When applying water to an exposure building, it is best to
    a. Apply water directly to the exposure building so that it washes down the side
    b. Apply water directly to the exposure building moving the stream in a rapid sequence of motions from side to side and up and down
    c. Discharge water in the form of a water curtain between the fire building and exposure building
    d. It is always best to attack the fire building.

12. An aerial ladder should not be _____ when a fire fighter is working on the ladder.
    a. Rotated
    b. Lowered or raised
    c. Extended or retracted
    d. Moved in any way

Control of Utilities

Chapter 12

Learning Objectives

- Realize that the knowledge of the company's *response area* is as important in utility control as it is in any other ladder company operation.

- Analyze each utility and the hazards associated with each service.

- Identify the proper technique to control heating, ventilation, and air conditioning systems.

- Examine the proper methods of controlling electrical systems.

- Examine the proper methods of controlling fuel services such as natural gas and oil systems.

- Examine the proper methods of controlling water distribution systems.

Fire fighters must have a good working knowledge of the utility services within a building, from a single family dwelling to a large commercial occupancy or high rise. During preincident planning, fire fighters should familiarize themselves with the locations of power switches and shutoff valves and determine whether each of these devices controls all or part of the utility system. Building utilities that must be controlled for the safety of fire fighters include the electrical service and heating, ventilation, and air conditioning services, as well as the water distribution service.

Controlling utilities can alleviate the potential of further damage to the building. It can also prevent injury or death to fire fighters and civilians working within the building or surrounding area. Controlling electrical utilities can prevent short circuits, arcing, and the possibility of electrocution caused by contact with charged electrical wires or equipment. Shutting off gas or oil-fired heating systems can eliminate the potential of fire and/or explosion caused by damaged, leaking, or ruptured fuel lines. Managing the water distribution system can prevent the additional weight and the accumulation of water throughout the building, especially in the basement. Controlling utilities within a building ensures a safer environment for fire fighters performing firefighting operations.

Utility services within a building can often contribute to problems during firefighting operations. These utilities include the electrical service, fuel services such as gas and oil, and the water distribution service. These utilities support the heating, ventilation, and air conditioning systems, as well as other vital design functions within a building. If these services are not controlled during an incident, they may add fuel to the fire, contribute to fire extension, create additional damage, and cause extremely hazardous conditions.

Air circulation, forced-air heating, and air conditioning systems can allow smoke, gases, and fire to be distributed through air ducts to uninvolved parts of a building, creating additional firefighting, rescue, and safety problems.

Oil and kerosene heating systems or their fuel supply systems can become damaged and leak fuel, thus feeding the fire. Gas heating systems or distribution lines can be damaged and cause gas leaks. Natural gas is much lighter than air and will rise and disperse in open areas but can accumulate in dead spaces. If the gas is leaking and reaches an ignition source, a violent explosion can occur. This could cause considerable injury

Key Points

Air circulation, forced-air heating, and air conditioning systems can allow smoke, gases, and fire to be distributed through air ducts to uninvolved parts of a building, creating additional firefighting, rescue, and safety problems.

Key Points

If gas has been ignited, it should be allowed to burn, protecting adjacent combustibles, until the source can be shut off.

Key Points

Distribution sources such as electrical vaults, manholes, substations, and transformer rooms should be off limits to fire fighters and handled by qualified utility company representatives.

or death to fire fighters as well as creating a difficult fire control scenario. If gas has been ignited, it should be allowed to burn, protecting adjacent combustibles, until the source can be shut off.

Electrical wiring that has been exposed, burned through, or otherwise damaged can cause a severe hazard if fire fighters should come in contact with it. Distribution sources such as electrical vaults, manholes, substations, and transformer rooms should be off limits to fire fighters and handled by qualified utility company representatives.

Water flowing from broken pipes increases the hazard to firefighting by adding to floor loads or by flooding lower levels where fire fighters may be working. The water also causes unnecessary damage to the structure and its contents. To minimize or eliminate such problems, it is important to monitor and control the utilities within the fire building. The task of utility control is usually assigned to ladder company personnel.

Preincident Planning for Utility Control

The type, size, placement, and complexity of the utilities in a particular structure usually depend on the occupancy. In commercial and industrial occupancies, they may also depend on the activities conducted within the building. Therefore, knowledge of the company's district is as important in utility control as it is in any other ladder company operations. Ladder companies should become well informed through preincident planning and inspection of any unusual, complex, or difficult

Key Points

To minimize or eliminate such problems, it is important to monitor and control the utilities within the fire building. The task of utility control is usually assigned to ladder company personnel.

During preincident planning, ladder crews should familiarize themselves with the locations of power switches and shutoff valves and determine whether each of these devices controls all or part of the utility system. This is especially important in the case of large, complex systems, such as those in manufacturing and processing plants, large commercial buildings, and apartment houses, schools, and hospitals. Such installations should be inspected as necessary to gain an understanding as to how the utilities are used. **Figure 12-1** shows an example of a preincident floor plan with the locations of the utilities and **shutoff valves** marked. **Figure 12-2** shows a sample of a guide to be used in conjunction with the plan to explain the symbols used in the plan.

In smaller structures, including one- and two-family dwellings, garden apartments, stores, garages, and small commercial buildings, neighborhood building patterns sometimes result in a common location for each utility control in each type of occupancy. Ladder company personnel should, when needed, have a good idea where these controls will be found.

Key Points

Ladder companies should become well informed through preincident planning and inspection of any unusual, complex, or difficult utility systems within their district.

Key Points

A fire department must know whom to call during an emergency incident. The utility company should be requested to respond as soon as possible, and command should request an estimated time of arrival.

Key Points

The type of controls installed in a utility system and the way they function depend on the type, size, and age of the system. Fire fighters must be familiar with the various types of operating controls.

---

utility systems within their district. They should also have a thorough understanding of the hazards associated with each type of utility.

A fire department must know whom to call during an emergency incident. The utility company should be requested to respond as soon as possible, and command should request an estimated time of arrival. Utility company representatives are valuable assets to the incident commander at an incident. They have the expertise to assist command in making informed decisions.

Operating Controls

Utilities are, for the most part, controlled through power switches, breakers, and shutoff valves. The type of controls installed in a utility system and the way they function depend on the type, size, and age of the system. Fire fighters must be familiar with the various types of operating controls. In particular, they should be fully aware of what these controls will do inside the building when activated or deactivated.

Key Points

During preincident planning, ladder crews should familiarize themselves with the locations of power switches and shutoff valves and should determine whether each of these devices controls all or part of the utility system.

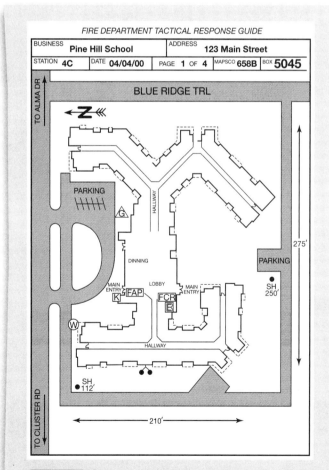

Figure 12-1   A preincident floor plan shows locations of utilities, including power switches and shut off valves.

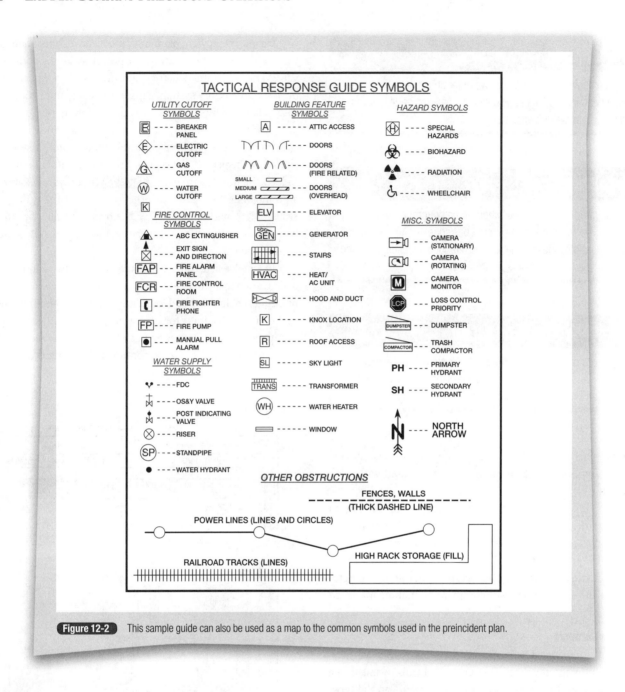

Figure 12-2 This sample guide can also be used as a map to the common symbols used in the preincident plan.

Building Codes

National and local building codes can be helpful in determining the locations of operating controls. Sometimes these codes require that the controls be placed in designated positions that

depend on the type, size, and layout of the building. Officers and fire fighters should be aware of the effect of these requirements on the structures in their district, whether they are small or large, residential, or commercial.

Forced-Air Systems

In a **forced-air system,** treated air is pushed through a series of ducts to the living and working areas of a structure. At the same time, air is drawn out of these areas through a second set of ducts. This "return air" is then treated and forced back to the living and working areas. Heating, cooling, filtering, dehumidifying, or a combination of these processes can treat the air that flows

---

Key Points

National and local building codes can be helpful in determining the locations of operating controls. Sometimes these codes require that the controls be placed in designated positions that depend on the type, size, and layout of the building.

Key Points

In a fire, the two sets of ducts form a system of channels for the horizontal and vertical spread of heat, smoke, gases, and flame.

through the system. It is forced through the ductwork by one or more fans or blowers **Figure 12-3** .

In a fire, the two sets of ducts form a system of channels for the horizontal and vertical spread of heat, smoke, gases, and flame. Dust and fibrous materials tend to collect along the bottom of the ducts, building up a fairly thick mat of combustible material over a period of years. If this material is ignited or heated sufficiently, the fire might flash through the entire ductwork system.

Forced-Air Blower

Action of the fan or **blower** compounds the problem in three ways. First, it increases the efficiency of the ducts as channels for fire spread. Second, it draws fire and heated gases into the ductwork through the return-air inlets, perhaps involving the ductwork. Third, it forces hot combustion products from involved areas to uninvolved areas through the air ducts. If these hot combustion products are discharged onto combustible material, there is a chance that they could be ignited. The fan then will repeat this cycle, forcing the heat from newly involved areas throughout the building.

For this reason, it is imperative to check the blower or blowers in an involved structure and to shut them down if they

Figure 12-3 Air return ducts could draw in smoke, heat, and gases, spreading them to other areas of the building.

Key Points

It is imperative to check the blower or blowers in an involved structure and to shut them down if they have not been shut down automatically or manually. Usually, it is best to simply shut down the entire heating or cooling system.

Key Points

Some larger systems also are equipped with dampers that close automatically to seal off portions of the ductwork and isolate the fire.

have not been shut down automatically or manually. Usually, it is best to simply shut down the entire heating or cooling system.

Some of the larger forced-air systems contain smoke detectors that monitor the air in the ductwork. If a detector senses smoke, it automatically shuts down the system. Fire fighters should be familiar with such systems and with the locations of their manual control switches. This type of system must also be checked to make sure that it has shut down, and if not, it should be shut down manually.

Dampers

Some larger systems also are equipped with **dampers** that close automatically to seal off portions of the ductwork and isolate the fire. The damper may be spring loaded and held open by a fusible link. When the heat of a fire melts the link, the spring pulls the damper shut.

Automatic dampers are usually located where ducts pierce fire walls or other major dividing walls. An easily opened inspection plate is located near each damper. The plate allows a quick check to determine whether the damper has closed. If a closed damper has been subjected to heat or fire, the duct on both sides of the damper must be inspected for fire spread.

These dampers are effective in stopping the spread of fire; however, experience has shown that large amounts of smoke and gases can travel past the dampers and spread through the building before the fusible links become hot enough to melt. This is why modern systems have dampers actuated by smoke

Key Points

If a closed damper has been subjected to heat or fire, the duct on both sides of the damper must be inspected for fire spread.

Key Points

Modern systems have dampers actuated by smoke detectors rather than by fusible links. Smaller systems, such as those in dwellings, do not normally have dampers. In any case, the forced-air system must be shut down during a working fire.

Key Points

When there is a fire in the living area, the increased temperature causes the burner thermostat to shut down the burner; however, the hot return air increases the plenum temperature, allowing the blower to continue to operate. This aids in the spread of fire, and thus, the system must be shut down.

detectors rather than by fusible links. Smaller systems, such as those in dwellings, do not normally have dampers. In any case, the forced-air system must be shut down during a working fire.

Heating Systems

The forced-air heating systems usually found in one- and two-family dwellings are designed so that a furnace produces the heat that is distributed throughout the house. The heat warms the air in a **plenum** located above the burner. A blower forces the warmed air through the ducts to the living areas and draws room air back to the plenum. Separate thermostats control the burner and blower. A thermostat in the living area turns the burner on to warm the dwelling and off when the desired temperature is reached. A thermostat in the warm-air plenum operates the blower as long as the air in the plenum is above a set temperature.

When there is a fire in the living area, the increased temperature causes the burner thermostat to shut down the burner; however, the hot return air increases the plenum temperature, allowing the blower to continue to operate. As noted previously, this action aids in the spread of fire, and thus, the system must be shut down.

Larger and more complex forced-air heating systems work on the same basic principle. Some, however, are divided into separate sections, or zones, each with its own blower. In larger structures, entire ceiling areas are sometimes used as part of the air-return system. In such cases, fire drawn into the system is not confined within a duct but can be easily spread over a large, unprotected area **Figure 12-4**.

In medium- and high-rise buildings, a single-duct circuit may serve several floors. Smoke, heat, and gases from a fire on

Key Points

In larger structures, entire ceiling areas are sometimes used as part of the air-return system. In such cases, fire drawn into the system is not confined within a duct but can be easily spread over a large, unprotected area.

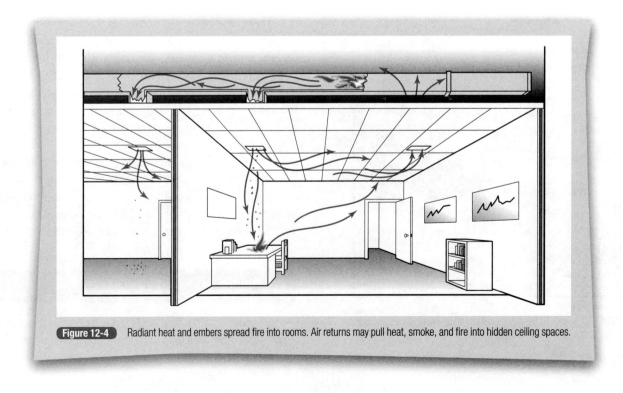

Figure 12-4 Radiant heat and embers spread fire into rooms. Air returns may pull heat, smoke, and fire into hidden ceiling spaces.

Key Points

In medium- and high-rise buildings, a single-duct circuit may serve several floors. Smoke, heat, and gases from a fire on one floor are easily carried to other floors, endangering people who would otherwise be relatively safe.

Key Points

As with a heating system, the operating blower of a cooling system can easily draw fire in through the return-air inlets and spread combustion products through the ductwork or ceiling area to the area it serves.

Key Points

When the outdoor temperature requires neither heating nor cooling indoors, forced-air blowers might be operating merely to circulate air through a structure. Again, fire fighters must shut down the operating system when a fire occurs.

Key Points

Some forced-air systems are designed so they also can be used to exhaust smoke from the building after the fire has been knocked down. Ladder companies should determine, through preincident planning and inspection, where such systems exist.

one floor are easily carried to other floors, endangering people who would otherwise be relatively safe. When this situation causes panic, it compounds the rescue problem.

Cooling Systems

The thermostat of a cooling (air conditioning) system keeps the system operating as long as the room temperature is above the set temperature. The heat from a fire will cause the system to run continuously in an attempt to reduce the room temperature to satisfy the thermostat setting. As with a heating system, the operating blower easily can draw fire in through the return-air inlets and spread combustion products through the ductwork or ceiling area to the area it serves.

When the outdoor temperature requires neither heating nor cooling indoors, forced-air blowers might be operating merely to circulate air through a structure. Again, fire fighters must shut down the operating system when a fire occurs.

Air Circulation Systems

In older buildings, including theaters, halls, and some other large structures, an obsolete type of air circulation system might still be in use. This system has large ducts that extend throughout the building. The ducts are connected to a huge air intake and a blower, usually located in the basement. In some cases, the system includes a primitive cooling setup in which cool water is circulated through large radiator-like units placed between the blower and the air intake. These units cool the air as it is blown into the system. Usually these systems vent through the regular ventilators in the building. In some cases, however, exhaust fans are used to provide faster air movement in hot weather. These venting fans are different from the venting systems installed over stages. The latter are used to vent the stage area and deter the spread of fire into the auditorium.

These air circulation systems almost always are controlled manually rather than automatically; therefore, they must be shut down manually, through a master power switch, when a working fire is encountered.

The problems that can be created by forced-air systems, and the extreme importance of shutting down these systems when a working fire is encountered has been discussed. Some forced-air systems are designed so that they also can be used to exhaust smoke from the building after the fire has been knocked down. Truck companies should determine, through preincident planning and inspection, where such systems exist.

Heating Units and Fuels

When a heating unit is involved with fire or in danger of becoming involved, both the unit and its fuel supply must be shut down. Heating fuels are, by their nature, combustible; it is important to make sure that they do not add to the firefighting problem.

Oil Burners

Oil burners are used to supply heat to many different occupancies—from small to large. The fuel oil is stored in a tank and pumped to a burner in the furnace. In the burner, the oil is vaporized, mixed with air, and burned. The burner, fuel pump, and heat-circulating device are operated electrically.

Small Systems   Small oil-burner systems, such as those found in one- and two-family dwellings, burn No. 2 fuel oil, which is similar to kerosene or diesel fuel. The storage tank can be located

Key Points

When a heating unit is involved in a fire or in danger of becoming involved, both the unit and its fuel supply must be shut down.

Key Points

Most building codes require that an emergency power shutoff switch be located at the top of the basement stairs if the unit is in the basement, at an outside basement entrance, or just inside an entrance to the building or the basement.

Key Points

If there is a break in the fuel line and the fuel tank is above the level of the burner, fuel will flow down toward the burner. The flow of fuel then can be stopped by closing the fuel shutoff valve located on the supply line right at the fuel tank.

in the basement or above ground, against the outside wall that is closest to the furnace. A fuel line, usually made of copper, carries fuel from the tank to the burner.

Most building codes require that an emergency power shutoff switch be located at the top of the basement stairs, if the unit is in the basement, at an outside basement entrance, or just inside an entrance to the building or the basement. The emergency switch is bright red. When it is turned off, it cuts all power to the unit, including the fuel pump, and thus, it ordinarily will stop the flow of fuel.

If, however, there is a break in the fuel line and the fuel tank is above the level of the burner, fuel will flow down toward the burner. The flow of fuel then can be stopped by closing the fuel shutoff valve located on the supply line right at the fuel tank. A supply line made of copper tubing can carefully be squeezed together with pliers to stop the flow of fuel if the valve will not work. If the supply line is made of heavy pipe, it will be necessary to cut the pipe and plug it.

Large Systems   Large oil-heating units, such as those in apartment houses, office buildings, factories, schools, and hospitals, burn a heavy oil, usually No. 6 fuel oil. Because the heavy oils do not ignite as readily as No. 2 oil, many of these systems include a device that preheats the fuel oil before it is burned.

The fuel is stored in underground tanks or in large aboveground tanks, usually located some distance from the building. The oil is pumped from the tanks to the preheaters

Key Points

The emergency power shutoff switches shut down the pumps, preheaters, and burners; this usually stops the flow of oil from the storage tanks.

Key Points

Aboveground and elevated tanks are especially prone to siphoning, and thus, the fuel flow should be checked after the power is turned off. If possible, the manual fuel shut-off valves also should be closed.

and then to the burners. The emergency power shutoff switches shut down the pumps, preheaters, and burners; this usually stops the flow of oil from the storage tanks; however, on occasion, a siphon is set up within the fuel line and fuel continues to flow even though the pumps have stopped. Aboveground and elevated tanks especially are prone to siphoning, and thus, the fuel flow should be checked after the power is turned off. If possible, the manual fuel shutoff valves also should be closed.

Some large oil-burning systems are sectionalized, with a separate subsystem serving each part of the structure. In such cases, it might be necessary to shut down only one subsystem to gain control of the fire; however, if there is any doubt, the entire system should be shut down. Building engineers can be of help in determining the proper course of action.

Kerosene Heaters and Stoves

Kerosene still is used as a fuel for both heating and cooking, but to a lesser extent than oil or gas. Kerosene heaters and cook stoves are found in rural areas and in poorer urban areas, where the cost of gas or oil is prohibitive or where older central-heating systems have deteriorated.

Kerosene heaters are self-contained units—that is, they are not connected into a separate heat distribution system as are oil burners. Some are designed to stand on the floor of a room, providing heat by convection out of the top and by a fan that blows heat out at floor level. Other kerosene heaters, known as floor furnaces, can be placed below the floor. A grate in the flooring above the unit allows the heat to rise into the living area by convection.

Kerosene cook stoves resemble standard gas or electric kitchen ranges. Both stoves and heaters can contain fuel tanks or can be fed fuel from an outdoor storage tank. Some are designed to operate from either type of fuel supply.

Kerosene-burning units do not usually have remote emergency switches; all of the controls are on the unit itself. This makes it necessary to cut off the electricity in the building to keep fans from operating if the unit cannot be reached during an emergency. In addition, the fuel shutoff valve at the outdoor tank must be closed to keep fuel from continuing to flow. Otherwise, the kerosene will feed the fire if the fuel supply line is damaged.

Key Points

Kerosene-burning units do not usually have remote emergency switches; all of the controls are on the unit itself. This makes it necessary to cut off the electricity in the building to keep fans from operating if the unit cannot be reached during an emergency. In addition, the fuel shutoff valve at the outdoor tank must be closed to keep fuel from continuing to flow. Otherwise, the kerosene will feed the fire if the fuel supply line is damaged.

Key Points

When fire has involved the area around a heater that has a self-contained fuel tank, the tank might explode and throw flaming fuel over the entire room. When it is known that such heaters are in use, fire fighters should approach involved rooms with extreme caution and with hose lines ready.

When fire has involved the area around a heater that has a self-contained fuel tank, the tank might explode and throw flaming fuel over the entire room. When it is known that such heaters are in use, fire fighters should approach involved rooms with extreme caution and with hose lines ready. If the tank has already exploded, standard firefighting operations should be used to bring the situation under control.

Gas Units

Gas, either piped (city gas) or bottled, is used for heating and cooking and in many manufacturing processes. Municipal gas is piped through meters to the point of use; it is usually a natural fuel but could be manufactured. Bottled gas is a liquefied petroleum product, such as propane, stored in heavy cylinders stored outdoors near the point of use. Both piped gas and bottled gas are highly combustible; a gas supply or outlet threatened by fire should be closed off to prevent ignition or explosion.

Municipal Gas   Many jurisdictions require that the main shutoff be located in plain sight outside of any building served by gas lines. The shutoff may be painted a distinctive color and marked "emergency gas shutoff." Outside main gas valves might also be located at or near outside meters or just outside the locations of indoor meters. When an outside main valve is not provided, the gas shutoff is located on or near the gas meter inside

Key Points

Both piped gas and bottled gas are highly combustible; a gas supply or outlet threatened by fire should be closed off to prevent ignition or explosion.

Key Points

The local gas company can be of great assistance in locating the outside shutoff or street valve.

the building. Unfortunately, this is the case when the meter is located in the basement. When a basement fire or a serious first floor fire makes it impossible to reach the meter, the street valve must be used to stop the flow of gas.

The local gas company can be of great assistance in locating the outside shutoff or street valve; however, the degree of cooperation depends on the individual gas company. The fire department should have a good relationship with all the utility companies in its jurisdiction. Utility company representatives usually have a good understanding of the conditions under which fire fighters work and are there to support the effort.

Gas Meters   The gas meter is usually the weakest link in the gas supply system. A meter located in an involved area will generally fail before the piping fails. When the meter fails, the escaping gas will ignite and form a burning jet. The escaping gas should be allowed to burn while the immediate area and exposures are protected. No attempt should be made to extinguish the gas with hose streams. Flames at a meter or at a damaged gas line should be extinguished only by shutting off the flow of gas.

If a gas fire is extinguished before the supply is shut off, the gas will probably re-ignite. Worse, it might collect and eventually explode, causing further damage to the gas supply system or the building and perhaps cause injury or death to occupants and fire fighters.

In many apartment buildings served by city gas, a separate meter is provided for each unit. Individual meters may be located in a central bank in the basement of the building or they may be outside the building. This arrangement allows fire fighters, or the local gas company, to shut off the gas selectively, instead of shutting off the flow to the entire building. More important, after the fire is extinguished, it allows undamaged gas appliances to be put back into operation while the supply of gas is held back from apartments with damaged appliances or supply lines. In addition, every gas appliance should have its own shutoff that could be closed if it was in an involved area. Gas company personnel should do the restoration of service.

Key Points

Flames at a meter or at a damaged gas line should be extinguished only by shutting off the flow of gas.

Key Points

In many apartment buildings served by city gas, a separate meter is provided for each unit. Individual meters may be located in a central bank in the basement of the building or they may be outside the building. This arrangement allows fire fighters, or the local gas company, to shut off the gas selectively, instead of shutting off the flow to the entire building.

When a fire involves the area in which a bank of gas meters is located, large amounts of gas can be released. The gas will add to the intensity of the fire or might collect and explode. The street valve should be closed under such conditions.

Ladder company personnel should be familiar with the locations of street valves, the main shutoff, inside and outside gas meters, and individual gas appliance shutoff valves. The apparatus should be equipped with any special tools or keys needed to close street valves or large valves used in industrial or large commercial installations. Again, the local gas company should be notified immediately to respond to the incident.

Bottled Gas Bottled gas for home use is usually stored in one or more outdoor cylinders located near or directly against the house. A copper line carries the fuel into the house to the appliances. Usually there are no meters, but a gauge and shutoff valve are located at the top of the cylinder or on the supply line if several tanks are being used. Closing the valve will shut off the flow of gas into the building. If for some reason the valve will not operate or will not completely shut off the flow, the copper line should be carefully squeezed shut with a pair of pliers **Figure 12-5**.

If a fire in the building endangers the cylinders, the cylinders and the exposed supply line should be cooled with a hose stream. Each cylinder has a relief valve or plug to relieve the cylinder

Figure 12-5 A copper line carries fuel from outdoor gas cylinders into the house to appliances. Closing the shutoff valve found at the top of the cylinders will shut off the flow of gas into the building.

pressure in case of fire. If this valve has been actuated, a jet of burning gas might issue from the cylinder. The gas should be allowed to burn while the cylinder and the area around it is cooled with a fog stream. Extinguishing the flame will not stop the flow of gas. The unburned gas might collect and re-ignite, causing a flash fire over a large area or an explosion.

The gas used in large industrial plants is stored in cylinders located above or below ground level. In most installations, buried piping carries the fuel to the plant. Shutoffs can be located at the cylinder and/or on the supply line outside the building. In addition, sectional valves might serve various parts of the plant, and a bank of remote control valves might be located at an engineer's station in the plant.

Ladder company personnel must be familiar with any such installations in their response area. Ladder crews must be able to locate and close off the appropriate valves quickly when the fire situation requires such action.

Key Points

When a fire involves the area in which a bank of gas meters is located, large amounts of gas can be released. The gas will add to the intensity of the fire or might collect and explode. The street valve should be closed under such conditions.

Key Points

Ladder company personnel should be familiar with the locations of street valves, the main shutoff, inside and outside gas meters, and individual gas appliance shutoff valves. The apparatus should be equipped with any special tools or keys needed to close street valves or large valves used in industrial or large commercial installations.

Key Points

If a fire in the building endangers the cylinders, they and the exposed supply line should be cooled with a hose stream. Each cylinder has a relief valve or plug to relieve the cylinder pressure in case of fire. If this valve has been actuated, a jet of burning gas might issue from the cylinder. The gas should be allowed to burn while the cylinder and the area around it is cooled with a fog stream.

Key Points

The gas used in large industrial plants is stored in cylinders located above or below ground level. In most installations, buried piping carries the fuel to the plant. Shutoffs can be located at the cylinder and/or on the supply line outside the building.

Electric Service

When a working fire is encountered, officers must consider shutting down the electric service in the involved area or in the entire structure. Fire extending into walls, ceilings, or floors could have burned the insulation off electric wires. In addition, electric fixtures might have been damaged and their wiring exposed. Fire fighters who contact the bare wiring are subject to serious injury or death. Hose streams conduct current and streams that hit exposed electrical equipment could cause electric shock. Ladder crews might strike exposed wiring with their tools when checking for fire extension, ventilating, or performing other duties.

The best way to prevent injury from electric shock is to shut down the electrical service in the fire building; however, in each situation the need for electricity during firefighting operations must be balanced against the dangers it presents. For example, suppose fire companies arrive at a working apartment house fire at night and find occupants attempting to leave the building. It would be senseless to immediately cut off the electrical supply and with it the lights being used by escaping occupants. The confusion and panic generated by the sudden darkness would far outweigh the benefits in terms of fireground safety.

The electrical service in a fire building can be used by fire fighters to power portable lights, fans, and electrically operated tools. The building lights can be of help in search and rescue, fire attack, and other operations; however, when the condition of the building indicates that electrical features might endanger fire fighters, the electricity should be shut off in the fire area or, if necessary, in the entire building. In the previous example, command might have the electricity shut off after it was ascertained that all occupants had escaped the fire building or the primary search has been completed.

Main Power Switches

The electrical service provided to a building depends on the type of occupancy. Dwellings usually require 110- or 220-volt electricity

> **Key Points**
>
> When a working fire is encountered, officers must consider shutting down the electric service in the involved area or in the entire structure.

> **Key Points**
>
> The best way to prevent injury from electric shock is to shut down the electrical service in the fire building; however, in each situation the need for electricity during firefighting operations must be balanced against the dangers it presents.

> **Key Points**
>
> The electrical service provided to a building depends on the type of occupancy. Dwellings usually require 110- or 220-volt electricity for lighting, heating, and electrical appliances. Industrial plants often require up to several thousand volts for manufacturing processes.

for lighting, heating, and electrical appliances. Industrial plants often require up to several thousand volts for manufacturing processes. The number, type, and locations of power switches depend on the type of electrical service provided.

Dwellings   The modern electrical service to a one- or two-family dwelling can be shut off completely by pulling, or removing, the electric meter. The utility company representative should accomplish this. Service to particular areas of the house can be shut off at the fuse box or circuit breaker box.

When the power switches are located in the fire area and cannot be reached, it might be necessary to cut the electric service line to the building. The utility company representative should perform this task. There are some fire departments that allow members to cut electric service lines to a building. This is not a recommended procedure and is best left to the utility company personnel who perform this work on a regular basis. They possess the expertise and proper equipment to perform this task safely.

Commercial and Industrial Structures   The situation is similar at larger buildings, although the voltages may be much higher. Usually a main power switch and several sectional switches allow the power to be cut either completely or selectively. Especially at night, it could be important to shut off the electricity in the fire area only so that the lights of nearby areas can be kept on during fire department operations.

> **Key Points**
>
> The modern electric service to a one- or two-family dwelling can be shut off completely by pulling, or removing, the electric meter. The utility company representative should accomplish this. Service to particular areas of the house can be shut off at the fuse box or circuit breaker box.

> **Key Points**
>
> When the power switches are located in the fire area and cannot be reached, it might be necessary to cut the electric service line to the building. The utility company representative should also perform this task.

Key Points

Fire fighters should seek the assistance of building engineers, maintenance personnel, and the utility company on the decision to shut down high-voltage electrical equipment.

Key Points

Fire fighters should not use elevators if at all possible in a building whose main electrical power has been lost.

In most commercial and industrial structures, electrical panels and lines that carry unusually high voltages are so marked. As noted earlier, fire fighters should not be allowed in areas where high-voltage equipment is located. This includes but is not limited to electrical vaults, manholes, substations, and transformer rooms. Fire fighters should seek the assistance of building engineers, maintenance personnel, and the utility company on the decision to shut down high-voltage electrical equipment.

Elevators

Elevators are usually powered by their own electrical source, separate from that for the remainder of the building. Fire fighters must follow standard operating guidelines and use extreme caution when using elevators in a fire building. If the power to the elevators is lost, usually a number of the elevators will be able to operate on emergency power. Fire fighters should not use elevators if at all possible in a building whose main electrical power has been lost.

Water Pipes

Although a damaged water system will not contribute to fire spread, it can add to operating problems or become a hazard to fire fighters. An uncontrolled flow of water will cause unnecessary damage to the building and its contents and add substantially to the floor load. The water also could flood lower levels where fire fighters are or may be working.

In cold climates, there is the added danger that the water will freeze, causing additional structural damage and difficult footing for fire fighters. Supply lines and drain pipes can freeze as the temperatures drop below freezing, causing the pipes to crack. These cracks will leak and become obvious as the temperature rises above freezing and thawing occurs.

Key Points

Although a damaged water system will not contribute to fire spread, it can add to operating problems or become a hazard to fire fighters.

Key Points

Ladder companies should maintain control of both the water system and the power system.

Key Points

In most communities, there is a pattern to the placement of the main water valves; ladder companies should be aware of both the pattern and the exceptions.

Key Points

In large structures, the water system often is divided into sections for water control during repairs; the locations of such sectional controls should be determined as part of preincident planning.

Furthermore, water can also cause electrocution if electrical wires energize it. Ladder companies should maintain control of both the water system and the power system. A damaged water system should be shut down as soon as possible to minimize problems due to excessive water or freezing and to reduce the extent of required property conservation operations.

Water Shutoff Valves

In communities that meter water, a street shutoff valve is usually located near the meter at the front of each building. There may be a shutoff valve in the street even where city water is not metered. In most communities, there is a pattern to the placement of these main water valves; ladder companies should be aware of both the pattern and the exceptions. Special keys might be needed to close street valves. The local water department representatives should be notified to respond.

In buildings with basements, the main shutoff valve is often located on the incoming water line just inside the basement wall. In buildings without basements, the shutoff valve should be located on the interior of an outside wall in some sort of utility area. Normally, there are separate shutoff valves at sinks, basins, toilets, and individual fixtures. Components that use water for any number of reasons should also have separate shutoffs.

In large structures, the water system often is divided into sections for water control during repairs; the locations of such sectional controls should be determined as part of preincident planning.

Boilers and Heating Units

When it is necessary to cut off the flow of water to industrial boilers or to water or steam heating units, the heating source also must be cut off; otherwise, a dangerous situation could be created. Water service should be restored to such units only as the manufacturer recommends.

Chief Concepts

- It is evident that ladder crews assigned to utility control at fires must have a thorough knowledge of the various utilities they are likely to encounter. Heating, ventilation, and air conditioning systems; electric, oil, kerosene, and gas burners; electrical systems; and water supply systems should be monitored and controlled to assist in fire control operations and to keep injury and damage to a minimum.

- Ladder company personnel should know how to locate and operate the different types of power switches and shutoff valves used in their district. They should know which structures contain sectionalized systems and be aware of what parts of the structure are served by each section. They should not hesitate to shut down all or part of a utility system when such action will result in safer or more efficient overall fireground operations. Command should rely on representatives from the different utility companies for their expertise and assistance.

Key Terms

Blower: A fan in a heating and air conditioning system (HVAC).

Dampers: Units that limit or close return air flow within a heating and air conditioning system.

Forced-air system: A heating and air conditioning system (HVAC) that is used for climate control in a building.

Plenum: A box or area above the burner in heating a system where air is heated before being distributed throughout the heating and air conditioning system (HVAC) system.

Shutoff valves: Valves used to turn off or shut off the fuel to the heating and air conditioning system (HVAC) system.

1. Natural gas is _____ air.
   a. Lighter than
   b. Heavier than
   c. The same molecular weight as
   d. The actual weight of natural gas varies depending on the petroleum fractions used to produce the gas.

2. If natural gas is burning at the gas meter just beyond the shutoff valve, the best tactic would be to
   a. Push the fire away from the valve using a full fog pattern while a fire fighter closes the valve.
   b. Extinguish the fire and then immediately close the valve.
   c. Extinguish the fire, and if the pipe is copper, immediately crimp the pipe to shut off the flow.
   d. Allow the fire to burn, protecting nearby combustibles until the source can be shut off.

3. If smoke is encountered at an electrical substation, which of the following would be the best course of action?
   a. Call the utility company. Force entry to determine the extent of the fire, and then control the fire using a nonconductive extinguishing agent.
   b. Call the utility company. Force entry, and attempt to shut down the equipment involved in fire from a remote location, then extinguish the fire when you are sure the equipment is de-energized using a nonconductive extinguishing agent.
   c. Call the utility company. Force entry. Attempt to shut down the equipment involved in the fire from a remote location, but do not attempt extinguishment.
   d. Call the utility company, but remain outside of the substation.

4. In regard to a forced-air system in a building, it is generally best to
   a. Limit or close the return air flow from the fire area using dampers or other built-in equipment to stop the circulation of smoke and other products of combustion.
   b. Limit or close heated (treated) air ducts using dampers or other provided means to stop the circulation of smoke and other products of combustion to areas outside the fire area.
   c. Close all dampers to limit smoke circulation.
   d. Shut down the heating and cooling system for the entire building.

5. Heating ventilation and air conditioning systems in large buildings often have automatic dampers that close, and this automatic closure is actuated when _____.
   a. Fusible links melt causing a spring-loaded mechanism to close the damper.
   b. Smoke detectors sense the products of combustion that cause the damper to close.
   c. Either fusible links or smoke detectors are used to actuate the closing device on the damper.
   d. Very few buildings built since 1965 are equipped with dampers; newer buildings use reversible air flow and positive pressure to control smoke movement.

6. Small oil-burning furnaces will be equipped with an emergency shutoff switch
   a. When this switch is used to shut down the pump, the flow of fuel is stopped, and any danger of a fuel leak is eliminated.
   b. When this switch is used to shut down the pump, fuel will continue to flow into the furnace, but at a lower rate.
   c. When this switch is used to shut down the pump, the flow of fuel is stopped unless there is a break in the pipe. If the pipe is ruptured, fuel will continue to flow when the oil tank is above the burner or rupture.
   d. This switch is used to shut down the blower, not the fuel supply.

7. Large oil-burning heating systems usually use No. 6 fuel oil, which is _____
   a. Classified as a highly flammable liquid similar to gasoline in physical properties
   b. Classified as a flammable liquid, but with a flash point higher than gasoline (85°F to 93°F)
   c. A combustible liquid equal to kerosene in terms of flammability with a flash point of 102°F to 105°F
   d. A combustible liquid that is much less flammable than kerosene

8. A fire threatens a bank of gas meters located in a basement of an apartment building and the utility company has been delayed; the incident commander has ordered the gas shut off. As the ladder company officer, what is the best method of shutting down the gas supply to the building?
   a. Close the distribution system valve serving the neighborhood.
   b. Close the street valve.
   c. Close the quarter-turn valve at the meter.
   d. All of the above are acceptable means of shutting down the gas supply.

9. Electrical service provided to a building is
   a. 12 volts
   b. 110 volts
   c. 220 volts
   d. Voltage is normally 110 or 220 in a dwelling, but varies by occupancy.

10. When cutting all electrical power to a dwelling
   a. Fire fighters should pull the meter.
   b. Fire fighters should cut the service entry.
   c. Fire fighters should only perform the tasks in **a** and **b** above if they are properly trained and have the equipment necessary to perform the task safely.
   d. Fire fighters should wait for the utility company and have the utility company cut power to the building.

11. In most cases, shutting down electrical service to a building will _____ the elevators.
   a. Not affect the power supply to the elevator
   b. Also shut down power to the elevator
   c. Will cause the elevators to operate in "emergency mode"
   d. No general statement can be made regarding the relationship between building power and elevator power.

Overhaul Operations

Chapter 13

Learning Objectives

- Recognize that overhaul operations can be a dangerous assignment for fire fighters.

- Realize the importance of an inspection of the building before a concentrated overhaul effort is begun.

- Understand the principle of controlling personnel during overhaul operations.

- Define the basic purpose of overhaul.

- Identify the hazards that may be encountered during overhaul operations.

- Identify the procedures to reduce the chances of injury to personnel while performing overhaul operations.

Overhaul operations conducted after the control of a fire incident can be a dangerous assignment for fire fighters. Those who were engaged in the control and extinguishment of the fire may be needed to perform overhaul activities. They may be tired and exhausted from strenuous physical activity put forth while in the performance of their duties during fire suppression operations. Before fire fighters are assigned to overhaul operations, they should be rested or additional personnel should be assigned before overhaul begins. A preinspection should be made to determine the overall condition of the building or area in which overhaul operations will take place. Hazards must be identified and corrections made to ensure the safety of the fire fighters. Command should evaluate the department's risks-versus-benefits guideline before committing fire fighters to an excessively hostile environment.

Fire fighters must be strictly supervised while working inside the building. Crew members must work in teams under the direct supervision of a company or sector officer. Fire fighters must be in full personal protective equipment, including SCBA, as carbon monoxide levels may be extremely high because of the smoldering effects of combustible materials found during overhaul. Safety officers also should be assigned during overhaul to monitor any hazardous conditions and to ensure that fire fighters are working safely. A **rapid intervention team** must be assigned in case an emergency situation occurs.

Overhaul is dangerous work, especially if the building has been damaged excessively. Those performing overhaul operations should pay particular attention to their surroundings and be in direct contact with command. Despite the efforts of the fire service, injury rates remain high during overhaul operations.

When the main body of fire has been brought under control, overhaul operations should commence. Overhaul is the hard, dirty job of searching for and extinguishing any remaining fire, sparks, or embers that are visible or in concealed spaces. The main purpose of overhaul is to make certain that no trace of fire remains to rekindle after fire fighters have left the scene. In addition, overhaul should ensure that the structure is left in a secure state, especially if it is to be partially or completely occupied soon after the fire. Because water damage can occur during overhaul, fire fighters must protect undamaged goods and furnishings as part of the property conservation efforts.

Command must consider the degree of cleanup operations of the premises. Debris should be removed, ensuring total extinguishment of the fire. Command must consider areas of

Key Points

Debris that could be vital in determining cause and origin should not be removed until an assessment of the area has been made.

the building that may need to be examined by a fire investigator. In this instance, debris that could be vital in determining cause, and origin should not be removed until an assessment of the area has been made. Debris that could be dangerous to fire fighters or to occupants who may continue to inhabit the building should be examined during overhaul and removed from the building if necessary. Fire fighters should perform overhaul operations to ensure that these tasks have been completed. Depending on the circumstances, building maintenance crews may be able to continue cleanup operations after the fire has been extinguished if command has determined that it is safe to do so.

At many working fires, overhaul may be the toughest firefighting assignment. It requires a knowledge of fire travel and building construction, an expertise in hand and power tools used during overhaul operations, and the stamina and muscle for prolonged periods of hard work immediately after fire control. Unfortunately, in many cases, fire fighters already exhausted from strenuous firefighting operations are immediately assigned to overhaul work.

Tired crews sometimes try to work too quickly and tend to take chances in an effort to get the job finished. This often results in mistakes and in injuries to fire fighters. During fire attack and related operations, ladder company members must work quickly and because the nature of the job may take calculated risks. Overhaul operations begins after the fire is placed under control, and there is less reason to rush or take chances. There is time enough for officers to evaluate the situation and develop an orderly and safe overhaul plan. In spite of this, the injury rate during overhaul operations is relatively high.

Three procedures can ensure the safety of fire fighters and reduce the frequency of injury during overhaul:
- Inspection of the area before initiating overhaul operations
- Assignment of rested or additional personnel to conduct overhaul operations
- Proper supervision of personnel during overhaul operations

These procedures are discussed in the first two sections of this chapter, whereas the remaining section deals with the overhaul operation itself.

Preinspection

An inspection of the building before a concentrated overhaul effort is begun should determine the overall condition of the area and ensure the safety of fire fighters assigned to that area.

Key Points

The main purpose of overhaul is to make certain that no trace of fire remains to rekindle after fire fighters have left the scene.

Key Points

An inspection of the building before a concentrated overhaul effort is begun should determine the overall condition of the area and ensure the safety of fire fighters assigned to that area.

The structural integrity of the building, or areas within, must be determined before fire fighters are allowed to continue working. The building may have been damaged because of the amount of water used during fire attack and extinguishment, especially when large volumes have been applied during defensive operations. There may be holes in the floors and roof. Stairways may have been compromised, and building components may be damaged or consumed. Areas of the building might be too hazardous to enter. The groups of fire fighters who will be assigned to overhaul operations may know little about such unsafe areas or their locations in the building. This could be compounded if they have not been involved in the firefighting effort in those areas or if they are additional crews specifically assigned for overall operations. Unless the building is inspected and unsafe areas are identified and marked off or rendered safe, there is bound to be confusion, accidents, and injuries.

The extent of this inspection, as well as the extent of the entire overhaul operation, will depend on the size of the fire. If the fire was small, perhaps confined to one room and its contents, an inspection of the fire room and the rooms around it will probably suffice. If a large part of the building were involved, the entire overhaul area should be cleared of fire fighters and inspected by appropriate officers with the skill and knowledge to make an intelligent observation as to the condition of the building. A building inspector also may be of assistance if available.

The purpose of the inspection is to make sure that the area in which overhaul operations will be conducted is safe. The inspecting officers should see to it that holes in the flooring are safely covered or barricaded, that unsafe stairways are marked or taped off, that structurally unsound areas of the building are

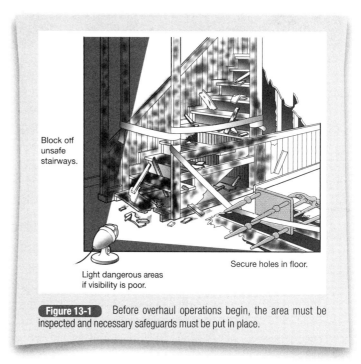

Block off unsafe stairways.

Light dangerous areas if visibility is poor.

Secure holes in floor.

Figure 13-1 Before overhaul operations begin, the area must be inspected and necessary safeguards must be put in place.

marked in an obvious way, and that portable lighting is placed in dark and dangerous locations, as shown in **Figure 13-1**.

When damage has been extensive at a night fire and the area cannot be properly lighted, the overhaul operation should be delayed until daylight if possible. The combination of extensive damage and poor visibility is extremely conducive to accidents. A fire watch should be established to extinguish any fire that might be rekindled, and sufficient ladder crews should remain on the scene to assist engine companies. Nonessential personnel should be placed in rehab or returned to service. When it is deemed safe to begin overhaul operations, sufficient personnel should be available to perform the task.

Personnel

If it is safe to begin overhaul operations soon after the fire is knocked down, fire fighters who are rested and in the best physical condition should be assigned to these duties. Those who have been fighting a fire for some time will be tired from physical exertion and worn down by heat, smoke, and the weight of their PPE and other equipment. Ladder crews who have been moving constantly from one assignment to another are generally not in condition to conduct a careful examination of the premises.

Key Points

The extent of this inspection, as well as the extent of the entire overhaul operation, will depend on the size of the fire.

Key Points

The purpose of the inspection is to make sure that the area in which overhaul operations will be conducted is safe.

Key Points

When damage has been extensive at a night fire and the area cannot be properly lighted, the overhaul operation should be delayed until daylight if possible.

Key Points

If it is safe to begin overhaul operations soon after the fire is knocked down, fire fighters who are rested and in the best physical condition should be assigned to these duties.

They deserve a rest before reassignment and should be placed in rehabilitation where they can be evaluated to determine if and when they can be reassigned, as shown in **Figure 13-2**. If they are needed for overhaul operations, they will be more alert after their rest and better able to perform assigned duties. Fully rested crews or additional fresh overhaul crews should be given work assignments before entering the building.

Overhaul operations should be assigned to fresher ladder company members who were not directly engaged in firefighting operations. Those who may have been assigned to less-strenuous outside activities could be used. As noted, fire fighters brought in specifically to relieve tired members could be assigned to overall operations. On call or volunteer departments, crews or individuals arriving on their own are candidates for overhaul duties.

Control of Personnel Movements

Fire fighters who have not been inside the fire building during the firefighting operations will not know which areas have been damaged or where the structure might be weak. The outside appearance of the building may give them little information about interior conditions. Even fire fighters who were working inside the building may have a different perspective after the fire has been knocked down and visibility has improved. It is therefore extremely important that such personnel be strictly supervised while working within the structure.

Key Points

Command will maintain control of the operation and assign fire fighters for overhaul operations as needed. No one should be permitted to enter the fire building without first reporting to an assigned officer.

The Incident Command System must continue to be used after the fire has been placed under control. Accountability remains in force. Command will maintain control of the operation and assign fire fighters for overhaul operations as needed. No one should be permitted to enter the fire building without first reporting to an assigned officer. Fire fighters entering the building without reporting may find themselves in serious trouble if no one knows that they are inside. Fire fighters shall wear full personal protective equipment, including SCBA, while working in the building **Figure 13-3**. Carbon monoxide levels will be extremely high because of the smoldering effects of combustible materials present during overhaul. Fire fighters performing overhaul shall work in teams under the direct supervision of a company or sector officer.

One or more safety officers should be assigned to monitor the operation. They should be aware of any hazardous condition that may contribute to fire fighter injuries and ensure that the

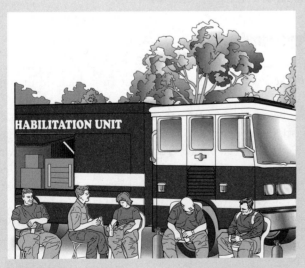

Figure 13-2 Crews who have worked during the initial attack phase should be allowed time to rest.

Figure 13-3 Fire fighters without protective clothing should not be allowed into the building.

fire fighters themselves are working safely and are physically capable of performing their duty.

A **fire fighter assist and safety team (FAST)** should be maintained throughout the overhaul operation, ensuring a fresh number of fire fighters who can be used with specific equipment if an emergency situation should occur.

The reason for these recommendations should be obvious. Although the fire might have been extinguished, there is a high probability that areas of the building are unsafe. Following these guidelines minimizes the chance of an accident, minimizes the likelihood of serious injury if there is an accident, and makes sure that help is available if needed.

Work Assignments

The basic duty of ladder crews in the overhaul operation is to locate and expose existing fire, embers, and sparks. This often requires opening various building components for examination or to further open spaces that were opened earlier to check for fire extension. Engine company personnel will advance and operate the hose lines needed to extinguish any small fires or embers that ladder company personnel may find.

Ladder crews should be assigned to the areas above, below, and adjacent to the fire area. It is in these locations that they will check for fire spread that has yet to be detected. Engine crews with hose lines should be stationed at the perimeter and called to these areas when a hot spot is discovered. In this way, personnel in the immediate area are kept to those members using hand and/or power tools to open concealed spaces. There is a good chance of fewer injuries to fire fighters in a less-congested area. Jobs that are neither ladder company work nor engine work must be shared. For example, if heavy items must be removed from an area, both engine and ladder crews are expected to lend a hand.

Procedure

The basic purpose of overhaul is to make sure the fire is completely out everywhere throughout the building. This means that any area that could possibly have been in contact with the fire or

Key Points

The basic duty of ladder crews in the overhaul operation is to locate and expose existing fire, embers, and sparks.

Key Points

Ladder crews should be assigned to the areas above, below, and adjacent to the fire area. It is in these locations that they will check for fire spread that has yet to be detected.

Key Points

The basic purpose of overhaul is to make sure the fire is completely extinguished everywhere throughout the building. Any area that could possibly have been in contact with the fire or intense heat should be checked, whether the building is fire resistant or not.

Key Points

The first step in the operation is to conduct an inspection of the structure to determine the stability of the building, especially when considerable damage has been done.

intense heat should be checked, whether the building is fire resistant or not. Often, building materials resist fire, but holes, shafts, cracks, voids, and other vertical and horizontal openings that are left during construction provide perfect channels for fire spread. Ladder crews should be assigned to check such building features, as well as the more common channels, and be prepared to open up any area suspected to contain fire.

As noted earlier, the first step in the operation is to conduct an inspection of the structure to determine the stability of the building, especially when considerable damage has been done. Consideration should be given to the building's age, type, construction, occupancy, and any other features that may affect its stability.

Figure 13-4 shows a dwelling in which a serious working fire was encountered.

Overhaul Hazards

Hazards encountered during overhaul operations may be numerous and varied. They could include

- Flooring weakened by the fire or damaged by firefighting operations.
- Openings made in the roof by fire fighters. The fire has also weakened the roof covering as well as the roof supports.
- Partially collapsed and weakened ceilings.
- The interior stairway having been partially consumed by the fire.
- Windows broken and shards of glass both inside and outside the building. Jagged edges of broken glass not being completely cleaned out of the window frames.
- Electrical wiring exposed and hanging throughout the building.
- Basement has a considerable amount of water accumulation because of the amount of water used to extinguish the fire.
- Hanging debris inside the building consisting of construction components with exposed nails and metal hangers.

Figure 13-4 Overhaul is dangerous work, especially if the building has been damaged excessively.

- Other hazards such as an air conditioning unit in broken windows, screens, and storm windows hanging from the building.
- Metal flashing and gutters hanging from the building.
- Exposed, damaged utility piping in the building feeding a natural gas stove and furnace.

Hazards encountered before overhaul operations must be confronted and corrections made to ensure the safety of fire fighters working within that environment.

Command should evaluate the department's risks-versus-benefits guidelines before committing fire fighters to an excessively hostile environment to perform overhaul operations.

Overhaul operations usually begin close to the areas in which firefighting operations ended, as the hose lines are already there. It is important, however, that other areas above, below, and adjacent to the fire area are also checked at the same time. There

should be an adequate number of hose lines properly positioned to support the operation.

Indications of Rekindling

Ladder companies assigned to overhaul duties should look for flames, smoke, a stronger-than-normal odor, and areas that flames have obviously touched. They should also look for vertical black streaks near baseboards and blistering and discoloration on walls that have not yet been checked, as shown in **Figure 13-5**. They should also listen for any crackling or unusual sounds and feel for heat that may indicate the presence of fire behind enclosed areas.

Concealed horizontal and vertical spaces should be checked, whether or not they were opened during fire attack and exposure protection operations. Portable lighting is of great help in examining concealed spaces or any areas that need to be opened up. They also show up smoke that otherwise might not be seen. The use of thermal imaging devices is of great assistance to fire fighters during overall, as they may indicate heat sources and where fires may not be extinguished.

A ceiling, floor, wall, or shaft that is opened for examination may show some sign of fire damage, usually in the form of blackened or charred surfaces. It then should be opened further

Key Points

Overhaul operations usually begin close to the areas in which firefighting operations ended, as the hose lines are already there.

Figure 13-5 During overhaul, beginning close to the area where firefighting operations ended, it is essential to look for signs of hidden fire.

until the full extent of the fire damage is indicated, usually by a clean area that was not touched by the fire. If flames, embers or smoke show when any space is opened, the area should be wet down and then further opened until the full extent of the fire damage is visible.

Areas of Possible Rekindling

Walls and Ceilings

If walls or ceilings have been in contact with fire and heat, they must be opened and checked for signs of live fire, as shown in **Figure 13-6**. If they have been partially opened in the course of

Figure 13-6 During overhaul, ceilings must be opened until a clean area is found.

firefighting operations, they should be opened further until the full extent of the fire is found. Engine crews should wet down any suspicious spots or areas with a light stream, and ladder crews should check adjoining ceilings and walls.

Ceiling spaces must be thoroughly examined because any fire there could be guided to wall spaces and then possibly through the building. Ceiling spaces should be checked with extra care so that damage will be minimized. Property-conservation efforts should be established to cover stock and furnishings in the area with salvage covers when ceilings are opened. If staffing is adequate and time allows, valuable contents may be removed from the area to a secure location. Every effort should be made to protect the building contents from further damage.

A check should be made to determine whether flames, embers, or sparks have been carried into interior walls or partitions. If so, there may be an indication of fire, smoke, or heat. Where there are such signs, an adequate opening should be made and a stream directed into the area to wet it down. Fire fighters should be able to detect signs of steam if a hot spot is encountered. These areas must be checked again before overhaul is completed **Figure 13-7**.

Above the Fire

Fire fighters working above the fire should remove baseboards for a positive check for fire travel through walls and partitions. Any doubtful area should be wet down and checked again later.

Older walls constructed of wood lath and plaster are more susceptible to hidden fire than newer walls made with metal studs and sheet-rock material. The old wood is extremely dry and easily ignited because of its rough finish. Older walls also

Figure 13-7   When fire is suspected behind walls, they should be opened until a clean area is found, and the area should be wet down with a stream if necessary.

have a wide inner space to accommodate wood joists. Walls of this type must be thoroughly inspected.

If a wall or ceiling contains insulation, both sides of the insulation must be checked. When the interior wall is opened, the paper or aluminum vapor barrier will be exposed. If no sign of fire is found, the other side can be checked by removing some of the insulating material. The insulation often can be removed in sections for a more thorough inspection.

In areas of a building exposed to fire containing cellulose blown-in insulation, it is imperative that the operation requires complete removal to ensure that the fire will not rekindle.

When ladder crews find that fire has entered into a ceiling space, they must assume that it has spread into the floor above. Fire fighters working on the floor above must check carefully to determine whether the fire has penetrated that space and whether there is further fire extension. A full examination of the area might require the removal of the flooring. This is especially important along walls and partitions and where vertical openings may pass through the floor. Again, if part of the flooring must be removed, ladder crews should take it up until a clean area shows the full extent of the fire. In order to hold damage to a minimum, the flooring should be removed as cleanly as possible. Cutting with power saws instead of using axes and prying tools may save time and conserve energy. The use of thermal imaging cameras is worthwhile to possibly reduce the need for additional damage to the structure.

In general, cutting during overhaul should be done with power saws rather than with axes and other hand tools. The saws cause much less vibration than hand tools and thus allow fire fighters a greater margin of safety. This is especially important in a building that has suffered structural damage and requires extensive overhaul work.

Vertical Passageways

When ladder crews suspect or find that fire has spread into a vertical passageway, it must be opened and checked **Figure 13-8** . Any natural openings into the passageway should be used for this purpose; otherwise, the shaft must be cut and checked for signs of fire, heat, and smoke. When necessary, a hose stream or streams should be directed into the shaft to wet down possible hot spots. In this case, the shaft opening might have to be enlarged to allow hose streams to be manipulated properly. In addition, ladder crews should be directed to check the top and bottom of the shaft for fire, embers, sparks, and heat.

Passageways that were opened for venting or fire control during firefighting operations must be thoroughly checked out at

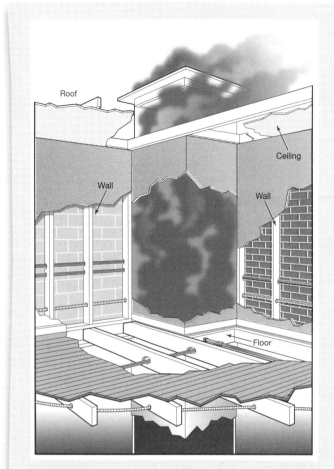

Figure 13-8   When a shaft has been the avenue of travel for fire, the areas above, below, and adjacent to the shaft must be opened.

this time. Even though the fire apparently has been extinguished, the intense heat that was confined within the vertical passageway might have rekindled a fire. For the same reason, anything in contact with these passageways such as floors, walls, attic spaces, or the roof must be thoroughly inspected. Floors, walls, and ceilings that abut them might have to be opened to assure a thorough check. If there is any sign of heat or fire extension, the floor, wall, or ceiling should be opened further until a clean, untouched area is found.

Cabinets and Compartments

The necessity of checking for fire extension around built-in cabinets and compartments located over or near fire areas was discussed in Chapter 6. If such cabinets have been subjected to fire or intense heat, they must be thoroughly checked during overhaul whether or not they were opened previously.

Window and Door Mouldings

When it is apparent that fire has involved window or door mouldings, ladder crews should remove the mouldings and check the concealed recesses for fire, as shown in **Figure 13-9** .

If the mouldings are in good condition, remove them without damage using an axe or pry tool so that they can be replaced when repairs are made.

If fire extension is found when the mouldings are removed, the walls or partitions must be opened to the end of the fire travel, and the area wet down as needed. Wainscoting is handled similarly; it is removed until a clean or unburned area is found, and the area is wet down if necessary.

Key Points

When it is apparent that fire has involved window or door mouldings, ladder crews should remove the mouldings and check the concealed recesses for fire.

Key Points

Even when a basement area has not been involved, check for fire that might have fallen from upper levels. This is especially important when it is known that fire traveled through vertical shafts that are open at the basement level.

Basement Areas

When fire has directly involved a basement or cellar, that area must be checked completely, the same as any other area of the fire building. Even when a basement area has not been involved, check for fire that might have fallen from upper levels. This is especially important when it is known that fire traveled through vertical shafts that are open at the basement level.

Walls Between Adjoining Structures

Party walls between adjoining occupancies also must be examined carefully. Both sides of such walls must be checked, with special attention given to the points at which joists from the two adjoining buildings overlap or abut. Fire may have traveled up into this area and continue to burn unless detected. Embers that easily lodge at these points and continue to smolder are difficult to find unless a careful check is made. Remember that if fire is suspected behind any hidden void, the area must be opened to determine the situation in that space.

Chemicals and Other Hazards

Fires involving chemicals, flammable liquids and/or solids, explosives, radiation, and other hazardous materials demand special consideration during both direct fire control activities and overhaul operations. Command shall determine whether overhaul operations are warranted and achievable under these conditions. Building personnel who are proficient in the chemical composition and use of hazardous materials can be of assistance in determining a course of action. In some cases, preincident planning surveys are of assistance as well as printed material, such as material safety data sheets or manifests. The type of material stored or in transit may be constantly changing, especially in general warehouses and freight depots. This situation may make positive identification difficult.

Key Points

Party walls between adjoining occupancies also must be examined carefully. Both sides of such walls must be checked, with special attention given to the points at which joists from the two adjoining buildings overlap or abut.

Key Points

Fires involving chemicals, flammable liquids and/or solids, explosives, radiation, and other hazardous materials demand special consideration both during direct fire-control activities as well as during overhaul operations.

Key Points

When dealing with a hazardous material, either the department's hazardous materials response team or a district or regional team should be notified and sent to the incident.

The characteristics of combustible materials and chemicals vary in many ways. Therefore, care must be exercised during all firefighting operations in an occupancy containing hazardous materials.

When dealing with a hazardous material, either the department's hazardous materials response team or a district or regional team should be notified and sent to the incident. Department standard operating guidelines for a hazardous materials response shall be followed and the ICS expanded depending on the scope and size of the incident.

If, after the fire has been brought under control, a decision has been made to conduct overhaul operations, it shall be conducted with fire fighters in PPE appropriate to the hazard. This may range from structural firefighting PPE to Level A encapsulated suits.

Those performing overhaul operations shall pay particular attention to their surrounding and be in direct contact with command. Fire fighters shall be continually monitored and upon leaving the hazardous zone shall be decontaminated along with their equipment before reporting to a medical sector for evaluation.

Searching for the Cause of the Fire

One important part of overhaul is the discovery of the cause of the fire and the possibility of the preservation of evidence in the event of arson **Figure 13-10**. Although not all fire fighters are arson investigators, they should be trained to look for signs of a deliberately set fire. When ladder company personnel conduct

Figure 13-10 Evidence of arson must be preserved.

Key Points

If, after the fire has been brought under control, a decision has been made to conduct overhaul operations, it shall be conducted with fire fighters in PPE appropriate to the hazard.

Key Points

Fire fighters shall be continually monitored and upon leaving the hazardous zone shall be decontaminated along with their equipment before reporting to a medical sector for evaluation.

overhaul operations without checking the fire building for such signs, the evidence might accidentally be thrown out of the building, buried, or washed away. It is important to assist in the determination of the cause of the fire. If there are illegal actions taken, this investigation should be turned over to the appropriate personnel in the fire department or law enforcement officials.

One indication of a deliberately set fire is its location. Such things as being in a bedroom or closet or in an unusual location may be a sign. Other things to check for include multiple fires, unusual odors such as gasoline or other flammable liquid, undue wood charring, uneven burning, holes made in walls and floors, heating equipment not in proper working condition, empty accelerant containers, residues of wax or paraffin, opened or removed service doors or panels to shafts, and inoperative sprinkler systems, fire doors, or other protective devices.

When a number of suspicious fires have occurred in an area, fire fighters should be notified of this activity. They should be especially alert for the deliberately set fire and command should call investigators to the fire scene when anything unusual is discovered. Although a particular fire department or company may operate at only one of the suspicious fires, investigators who have been to several can look for similarities among the fires that might be used to establish a case of arson.

Key Points

Although not all fire fighters are arson investigators, they should be trained to look for signs of a deliberately set fire.

Key Points

When a number of suspicious fires have occurred in an area, fire fighters should be notified of this activity.

Restoration and Protection

When overhaul operations have been completed, the building and its contents should be restored and protected, as much as possible, from the elements and from vandalism. In most instances, the police department, owner, or authority having jurisdiction would take over control of the building.

The Building

If vertical ventilation was required, skylights might have been removed, scuttles cut open, or holes cut in the roof. These openings should be closed or covered before the fire department leaves the scene.

A skylight that was removed without damaging the glass or frame can be replaced in its original position. Cover skylights with a salvage cover or plastic sheet if a good seal cannot be established at the roofline.

Large holes in a flat roof can be covered with salvage covers or plastic sheets. First, a ridge board and other boards as necessary should be placed to allow water to run off the roof and to keep the cover material from falling or sagging into the opening. The roofing material can be propped up if necessary to prevent water from running into the opening. A salvage cover should not be nailed to the roof. Instead, each end should be wrapped around a board that is longer than the cover and nails

Key Points

If vertical ventilation was required, skylights might have been removed, scuttles cut open, or holes cut in the roof. These openings should be closed or covered before the fire department leaves the scene.

driven through the protruding board ends into the roof. Both ends of the cover should be secured in this way for complete protection.

Roof drains on flat roofs should be cleared of debris and kept open; otherwise, in a heavy rain, the roof water level will rise, and water will seep into the roof holes. Openings in pitched roofs can be covered with salvage covers, tarpaper, or plastic, depending on their size. The roofing material above the hole should overlap the top of the cover, which should be held in place with wood strapping or slats.

Furnishings and Stock

Items removed from the building during the course of firefighting, property conservation operations, or overhaul should be returned to the building if possible. If the items must remain outside, place salvage covers over them and tie them down securely. The fire department should arrange for the security of these items as well as the covers and have the owners relieve the fire department of responsibility in writing.

When items that are returned to the building could be subjected to damage from dripping water or weather because of the condition of the building, these items also should be covered. If the building cannot be locked securely, make arrangements for protection or for a release from responsibility.

Key Points

Items removed from the building during the course of firefighting, property-conservation operations, or overhaul should be returned to the building if possible. If the items must remain outside, place salvage covers over them and tie them down securely.

Key Points

When items that are returned to the building could be subjected to damage from dripping water or weather because of the condition of the building, these items also should be covered.

Chief Concepts

- The main purpose of overhaul is to ensure that the fire is completely out before the departure of the fire department. For this, the fire structure must be thoroughly examined for fire, smoke, embers, sparks, and hot spots. As in other firefighting operations, coordination between ladder and engine crews is essential. Ladder crews must find and expose any ignition source that could rekindle another fire. Engine crews must wet down any area that shows signs of fire, heat, or smoke. These operations should be performed as carefully and as safely as possible, with a minimum of additional damage to the building and its contents.
- To reduce the chance of injuries to operating personnel during overhaul the following information should be noted:
  - Conduct an inspection of the building before overhaul begins.
  - Determine whether it is safe to perform overhaul operations.
  - Identify and mark hazardous areas.
  - If warranted, barricade and/or restrict personnel from hazardous areas in the building.
  - Use portable lighting in areas that lack sufficient lighting.
  - Assign fresh personnel to overhaul duties. Continually monitor their activities and replace them when they become tired.
  - See that overhaul personnel work as companies or in sectors under the direct supervision of an officer.
  - A safety officer should be assigned during overhaul operations.
  - Maintain FAST personnel during overhaul operations.
  - Observe special precautions in structures in which hazardous materials are stored.

Key Terms

Fire fighter assist and safety team (FAST): Similar to a rapid intervention team in makeup but is designed to replace and support personnel during overhaul operations.

Rapid intervention team: A minimum of two fully equipped personnel on site, in a ready state, for immediate rescue of injured or trapped fire fighters.

Fire Fighter in Action

1. When there has been extensive damage to a building during a night fire, it is best to
   a. Overhaul using portable lighting; if the entire building cannot be lighted at once, light each room as you proceed from the immediate fire area, outward, then upward.
   b. Overhaul using whatever lighting is available after a short break during which a careful inspection of the building is made and ventilation is provided.
   c. Overhaul immediately so that companies can return to service.
   d. If possible, establish a fire watch and wait until morning to overhaul.

2. During the overhaul phase
   a. Fire fighters should be allowed to remove those articles of personal protective clothing that are no longer needed as determined by the overhaul inspection.
   b. Fire fighters should remain in full personal protective clothing but may remove their SCBA.
   c. Fire fighters should remain in full personal protective clothing, including the SCBA.
   d. The IC should determine the correct level of protective clothing based on atmospheric testing and the overhaul inspection.

3. Overhaul occurs during the nonemergency phase of the operation; therefore,
   a. A more relaxed command system is in order, but the accountability system should remain in effect.
   b. The formal accountability process can be eliminated and a more relaxed command system established.
   c. The formal command and accountability systems can be eliminated, but the buddy system must be enforced.
   d. The ICS and accountability system must remain in effect during the entire operation.

4. When conducting overhaul operations
   a. Engine and ladder crews should work together in each room—the ladder crew gaining access to concealed spaces and the engine crew in the room with a charged hose line.
   b. Ladder crews should work in each room checking for fire extension and opening spaces as needed with the engine crew standing by near the perimeter, ready to extinguish any embers or small fires the ladder crew finds.
   c. Engine crews should remove heavy objects and debris while the ladder crew opens walls and other concealed spaces.
   d. Engine and ladder crews should be assigned to various areas, each accessing concealed spaces and each equipped with a hose line.

5. When opening a ceiling to check for fire extension
   a. Pull the entire ceiling in any room showing signs of fire entering the ceiling space.
   b. Continue opening the ceiling until there is no sign of fire damage.
   c. Open small inspection holes in several areas until there is no sign of fire damage.
   d. Open the ceiling just enough to operate a hose line into the false space, then operate the nozzle on a full fog pattern for at least 5 minutes.

6. Walls constructed of wood lath and plaster
   a. Are more susceptible to hidden fire than walls constructed using wooden studs covered with drywall
   b. Are more susceptible to hidden fire than walls constructed using metal studs covered with drywall
   c. Are more susceptible to hidden fire than walls constructed using metal or wooden studs covered with drywall
   d. Are less susceptible to hidden fire than walls constructed using metal or wooden studs covered with drywall

7. Which of the following would be an indicator of an intentionally set fire?
   a. A point of origin at the bottom of a stairwell
   b. Multiple points of origin
   c. Holes in floors
   d. Any of the above

8. When covering a roof opening during termination operations:
   a. Use an old salvage; cover and nail it to the roof. If two covers are needed, overlap the covers placing the top cover over the lower cover.
   b. Use an old salvage cover, slats, and/or boards; wrap the salvage cover over the edges of the slats and/or boards, and then nail the cover and boards to the roof driving the nail through both the cover and boards/slats. If two covers are needed, overlap the covers placing the top cover over the lower cover rolling the cover over the boards/slats.
   c. Roll a salvage cover over boards and/or slats that are longer than the cover. Roll the cover around the boards so ends of the board protrude beyond the cover and nail the ends of the board (not the cover) to the roof.
   d. Do not use salvage covers to cover roof openings.

Appendix

Imperial and Metric Conversions

Table A-1 Length

1 inch = 0.08333 foot, 1,000 mils, 25.40 millimeters

1 foot = 0.3333 yard, 12 inches, 0.3048 meter, 304.8 millimeters

1 yard = 3 feet, 36 inches, 0.9144 meter

1 rod = 16.5 feet, 5.5 yards, 5.029 meters

1 mile = (U.S. and British) = 5,280 feet, 1.609 kilometers,
0.8684 nautical mile

1 millimeter = 0.03937 inch, 39.37 mils, 0.001 meter,
0.1 centimeter, 100 microns

1 meter = 1.094 yards, 3.281 feet, 39.37 inches, 1,000 millimeters

1 kilometer = 0.6214 mile, 1.094 yards, 3,281 feet, 1,000 meters

1 nautical mile = 1.152 miles (statute), 1.853 kilometers

1 micron = 0.03937 mil, 0.00003937 inch

1 mil = 0.001 inch, 0.0254 millimeters, 25.40 microns

1 degree = 1/360 circumference of a circle, 60 minutes,
3,600 seconds

1 minute = 1/60 degree, 60 seconds

1 second = 1/60 minute, 1/3600 degree

Table A-2 Area

1 square inch = 0.006944 square foot, 1,273,000 circular mils,
645.2 square millimeters

1 square foot = 0.1111 square yard, 144 square inches,
0.09290 square meter, 92,900 square millimeters

1 square yard = 9 square feet, 1,296 square inches, 0.8361 square
meter

1 acre = 43,560 square feet, 4,840 square yards, 0.001563 square
mile, 4,047 square meters, 160 square rods

1 square mile = 640 acres, 102,400 square rods, 3,097,600 square
yards, 2.590 square kilometers

1 square millimeter = 0.001550 square inch, 1.974 circular mils

1 square meter = 1.196 square yards, 10.76 square feet, 1,550 square
inches, 1,000,000 square millimeters

1 square kilometer = 0.3861 square mile, 247.1 acres,
1,196,000 square yards, 1,000,000 square meters

1 circular mil = 0.7854 square mil, 0.0005067 square millimeter,
0.0000007854 square inch

Table A-3 Volume (Capacity)

1 fluid ounce = 1.805 cubic inches, 29.57 milliliters, 0.03125 quarts
(U.S.) liquid measure

1 cubic inch = 0.5541 fluid ounce, 16.39 milliliters

1 cubic foot = 7.481 gallons (U.S.), 6.229 gallons (British),
1,728 cubic inches, 0.02832 cubic meter, 28.32 liters

1 cubic yard = 27 cubic feet, 46,656 cubic inches, 0.7646 cubic
meter, 746.6 liters, 202.2 gallons (U.S.), 168.4 gallons (British)

1 gill = 0.03125 gallon, 0.125 quart, 4 ounces, 7.219 cubic inches,
118.3 milliliters

1 pint = 0.01671 cubic foot, 28.88 cubic inches, 0.125 gallon, 4 gills,
16 fluid ounces, 473.2 milliliters

1 quart = 2 pints, 32 fluid ounces, 0.9464 liter, 946.4 milliliters,
8 gills, 57.75 cubic inches

1 U.S. gallon = 4 quarts, 128 fluid ounces, 231.0 cubic inches,
0.1337 cubic foot, 3.785 liters (cubic decimeters), 3,785
milliliters, 0.8327 Imperial gallon

1 Imperial (British and Canadian) gallon = 1.201 U.S. gallons,
0.1605 cubic foot, 277.3 cubic inches, 4.546 liters (cubic
decimeters), 4,546 milliliters

1 U.S. bushel = 2,150 cubic inches, 0.9694 British bushel,
35.24 liters

1 barrel (U.S. liquid) = 31.5 gallons (various industries have special
definitions of a barrel)

1 barrel (petroleum) = 42.0 gallons

1 millimeter = 0.03381 fluid ounce, 0.06102 cubic inch, 0.001 liter

1 liter (cubic decimeter) = 0.2642 gallon, 0.03532 cubic foot,
1.057 quarts, 33.81 fluid ounces, 61.03 cubic inches, 1,000
milliliters

1 cubic meter (kiloliter) = 1.308 cubic yards, 35.32 cubic feet,
264.2 gallons, 1,000 liters

1 cord = 128 cubic feet, 8 feet × 4 feet × 4 feet, 3.625 cubic meters

Table A-4 Weight

1 grain = 0.0001428 pound

1 ounce (avoirdupois) = 0.06250 pound (avoirdupois), 28.35 grams, 437.5 grains

1 pound (avoirdupois) = the mass of 27.69 cubic inches of water weighed in air at 4°C (39.2°F) and 760 millimeters of mercury (atmospheric pressure), 16 ounces (avoirdupois), 0.4536 kilogram, 453.6 grams, 7,000 grains

1 long ton (U.S. and British) = 1.120 short tons, 2,240 pounds, 1.016 metric tons, 1016 kilograms

1 short ton (U.S. and British) = 0.8929 long ton, 2,000 pounds, 0.9072 metric ton, 907.2 kilograms

1 milligram = 0.001 gram, 0.000002205 pound (avoirdupois)

1 gram = 0.002205 pound (avoirdupois), 0.03527 ounce, 0.001 kilogram, 15.43 grains

1 kilogram = the mass of 1 liter of water in air at 4°C and 760 millimeters of mercury (atmospheric pressure), 2.205 pounds (avoirdupois), 35.27 ounces (avoirdupois), 1,000 grams

1 metric ton = 0.9842 long ton, 1.1023 short tons, 2,205 pounds, 1,000 kilograms

Table A-5 Density

1 gram per millimeter = 0.03613 pound per cubic inch, 8,345 pounds per gallon, 62.43 pounds per cubic foot, 998.9 ounces per cubic foot

Mercury at 0°C = 0.1360 grams per millimeter basic value used in expressing pressures in terms of columns of mercury

1 pound per cubic foot = 16.02 kilograms per cubic meter

1 pound per gallon = 0.1198 gram per millimeter

Table A-6 Flow

1 cubic foot per minute = 0.1247 gallon per second, 0.4720 liter per second, 472.0 milliliters per second = 0.028 m^3/min, lcfm/ft^2 = 0.305 m^3/min/m^2

1 gallon per minute = 0.06308 liter per second, 1,440 gallons per day, 0.002228 cubic foot per second

1 gallon per minute per square foot = 40.746 mm/min, 40.746 l/min · m^2

1 liter per second = 2.119 cubic feet per minute, 15.85 gallons (U.S.) per minute

1 liter per minute = 0.0005885 cubic foot per second, 0.004403 gallon per second

Table A-7 Pressure

1 atmosphere = pressure exerted by 760 millimeters of mercury of standard density at 0°C, 14.70 pounds per square inch, 29.92 inches of mercury at 32°F, 33.90 feet of water at 39.2°F, 101.3 kilopascal

1 millimeter of mercury (at 0°C) = 0.001316 atmosphere, 0.01934 pound per square inch, 0.04460 foot of water (4°C or 39.2°F), 0.0193 pound per square inch, 0.1333 kilopascal

1 inch of water (at 39.2°F) = 0.00246 atmosphere, 0.0361 pound per square inch, 0.0736 inch of mercury (at 32°F), 0.2491 kilopascal

1 foot of water (at 39.2°F) = 0.02950 atmosphere, 0.4335 pound per square inch, 0.8827 inch of mercury (at 32°F), 22.42 millimeters of mercury, 2.989 kilopascal

1 inch of mercury (at 32°F) = 0.03342 atmosphere, 0.4912 pound per square inch, 1.133 feet of water, 13.60 inches of water (at 39.2°F), 3.386 kilopascal

1 millibar (1/1000 bar) = 0.02953 inch of mercury. A bar is the pressure exerted by a force of one million dynes on a square centimeter of surface

1 pound per square inch = 0.06805 atmosphere, 2.036 inches of mercury, 2.307 feet of water, 51.72 millimeters of mercury, 27.67 inches of water (at 39.2°F), 144 pounds per square foot, 2,304 ounces per square foot, 6.895 kilopascal

1 pound per square foot = 0.00047 atmosphere, 0.00694 pound per square inch, 0.0160 foot of water, 0.391 millimeter of mercury, 0.04788 kilopascal

Absolute pressure = the sum of the gage pressure and the barometric pressure

1 ton (short) per square foot = 0.9451 atmosphere, 13.89 pounds per square inch, 9,765 kilograms per square meter

Table A-8 Temperature

Temperature Celsius = 5/9 (temperature Fahrenheit – 32°)

Temperature Fahrenheit = 9/5 × temperature Celsius + 32°

Rankine (Fahrenheit absolute) = temperature Fahrenheit + 459.67°

Kelvin (Celsius absolute) = temperature Celsius + 273.15°

Freezing point of water: Celsius = 0°; Fahrenheit = 32°

Boiling point of water: Celsius = 100°; Fahrenheit = 212°

Absolute zero: Celsius = –273.15°; Fahrenheit = –459.67°

Table A-9 Sprinkler Discharge

1 gallon per minute per square foot (gpm/ft^2) = 40.75 liters per minute per square meter (Lpm/m^2) = 40.75 millimeters per minute (mm/min)

Glossary

Aerial device: Any ladder, platform, pumper, or other apparatus designed and operated to support water, rescue, or fire fighter operations.

Aerial ladder: A power-operated ladder permanently mounted on a piece of apparatus.

Backdraft: The sudden explosive ignition of fire gases when oxygen is introduced into a superheated space previously deprived of oxygen.

Blower: A fan in a heating and air conditioning system (HVAC).

Catchalls: Salvage covers rigged as basins to catch and hold water dripping from overhead, as from a ceiling.

Combination ladders: Can be used as a single ladder, an extension ladder, and an A-frame.

Conduction: The travel of heat through a solid body. Although normally the least of the problems at a fire, the chance of fire travel by conduction should not be overlooked. Conduction can take heat through walls and floors by way of pipes, metal girders, and joists and can cause heat to pass through solid masonry walls.

Conduits: Salvage covers folded for use as conduits to direct accumulated water to stairways and then down the stairways and out through exterior doorways.

Convection: The travel of heat through the motion of heated matter—that is, through the motion of smoke, hot air, heated gases, and flying embers.

Cutting tools: Tools that are designed to cut into metal or wood.

Dampers: Units that limit or close return air flow within a heating and air conditioning system.

Elevating platform: A self-supporting, turntable-mounted device consisting of a personnel-carrying platform attached to the uppermost boom of a series of power-operated booms that articulate and/or telescope, sometimes arranged to provide the continuous egress capabilities of an aerial ladder.

Exposure fire: A fire that spreads from one structure to another or from one independent part of a building to another part from the original fire.

Exposure protection: The protection of other exposed structures by water streams.

Extension ladders: Adjustable in height and has a base ladder and one or more fly sections that travel in guides allowing for extension of the fly section.

Fire fighter assist and safety team (FAST): Similar to a rapid intervention team in makeup but is designed to replace and support personnel during overhaul operations.

Flashover: The ignition of combustibles in an area heated by convection, radiation, or a combination of the two. The action can be a sudden ignition in a particular location, followed by rapid spread, or a "flash" of the entire area.

Folding ladders: Single ladder that has hinged rungs; this allows the ladder to be folded so that the beams rest against each other.

Forced-air system: A heating and air conditioning system (HVAC) that is used for climate control in a building.

Fox lock: A device with from two to eight bars that hold the door closed from the inside.

Hand tools: Usually categorized as cutting, prying, or push/pull devices. These tools can be used to break locks, to force exterior and interior doors, to open sidewalk doors and grates, and generally to provide entry into a building.

Immediate rescue: This action must be attempted in extreme cases, such as when an arriving company finds occupants about to leave the building from upper floors by jumping because of deteriorating conditions. In such situations, ladder company crews must delay all other operations in favor of raising ladders.

Incident commander: Officer who is in charge of the incident and who develops a strategic plan for performing tasks during fireground operations.

Incident management system (IMS): The combination of facilities, equipment, personnel, procedures, and communications under a standard organizational structure to manage assigned resources effectively to accomplish stated objectives for an incident.

Light shafts: Four-sided shafts with skylight-type coverings.

Lock pullers: Such as the K-tool, are designed to remove cylinder locks.

Mushrooming: The shape of combustion products on the fire floor when they have accumulated under the roof and are forced to move horizontally.

Natural roof features: Skylights, scuttles, and penthouses and vertical shafts are examples of natural roof features that can be used to ventilate the building.

Penthouse: A small hut-like enclosure built over a stairwell that allows the stairs to extend up to the roof level and is tall enough to permit a person to walk onto the roof.

Personal alert safety system: Device worn by a fire fighter that sounds an alarm if the fire fighter is motionless for a period of time.

Plenum: A box or area above the burner in heating a system where air is heated before being distributed throughout the heating and air conditioning system (HVAC) system.

Pompier ladders: Consist of a single-beam ladder with rungs protruding from either side. A large hook at the top is placed on the sill in an open window or other opening. Used to climb from floor to floor by way of the window openings.

Power tools: Electricity, gasoline engines, or air or hydraulic pressure powers these tools.

Property conservation: Procedures and operations used to protect and reduce damage to the building and its contents from the effects of smoke, heat, fire, and water damage during firefighting operations. It is the third tactical priority behind life safety and extinguishment.

Prying tools: Tools designed to provide a mechanical advantage using leverage to force objects, in most cases, to open or break.

Quint: A type of fire apparatus with a permanently mounted fire pump, a water tank, a hose storage area, an aerial ladder, or elevating platform with a permanently mounted waterway and a complement of ground ladders.

Radiation: The travel of heat through space; no material substance is required. Pure heat travels away from the fire area in the same way as light—that is, in straight lines. It is unaffected by wind and, unless blocked, radiates evenly in all directions.

Rapid intervention team: A minimum of two fully equipped personnel on site, in a ready state, for immediate rescue of injured or trapped fire fighters.

Roof ladders: Designed in one section only, equipped with folding hooks at the tip to provide an anchoring mechanism when the ladder is placed over the peak or ridgepole of a building. Designed to lie flat on a roof as a work platform.

Salvage covers: Large sheets of waterproof material that are available in several materials, dimensions, and shapes. Some are fire resistant, and some are not.

Scuttle: Construction feature of a roof that allows access to the roof from inside the building.

Shutoff valves: Valves used to turn off or shut off the fuel to the heating and air conditioning system (HVAC) system.

Striking tools: Tools designed to strike other tools or objects such as walls, doors, or floors.

Telescoping waterway: Extending from the base of the bed ladder to the master stream appliance located at the tip of the fly section; allows the master stream appliance to be operated from any height from the tip of the bed section to the tip of the fully extended fly section.

Trench cut: A cut that is made from bearing wall to bearing wall to prevent horizontal fire spread in a building.

Water tower: An aerial device consisting of permanently mounted power-operated booms and a waterway designed to supply a large capacity mobile elevated water stream.

Index

Photo Credits

Chapter 1
Opener © Nathan DeMarse/ShutterStock, Inc.

Chapter 2
Opener © John Wollwerth/ShutterStock, Inc.

Chapter 4
Opener © Glen E. Ellman

Chapter 5
Opener © Larry St. Pierre/ShutterStock, Inc.

Chapter 8
Opener © Nathan DeMarse/ShutterStock, Inc.

Chapter 11
Opener © Nathan DeMarse/ShutterStock, Inc.

Chapter 12
Opener courtesy of Barry Hyvarinen, Massachusetts Fire Academy

Chapter 13
Opener © Glen E. Ellman

Fire Fighter in Action image © Jack Dagley Photography/ShutterStock, Inc.

Unless otherwise indicated, photographs are under copyright of Jones and Bartlett Publishers and were photographed by the Maryland Institute of Emergency Medical Services Systems.